LTE and IoT
Systems and Signal Quality

Sivannarayana Nagireddi, PhD

Printed in United States of America`

ISBN: 1534630686
ISBN 13: 978-1534630680

Library of Congress Control Number: 2016909686

Contents

Acknowledgments x

About the Author xi

Preface xii

1 LTE SYSTEM 1

1.1   LTE UE Architecture and Protocol Layers 3

1.2   LTE UE Uplink 4

1.3   LTE UE Downlink 6

1.4   Summary of Call-Flow Steps 7

1.5   LTE Frame, Subframe, Slot, and Symbols 7

1.6   LTE BW and Time-Frequency Parameters 9

1.7   Adaptive Modulation and Coding (AMC) 10

1.8   Frequency Bands 11

1.9   Frequency-Division Duplexing 11

1.10   Time-Division Duplexing 13

1.11   TDD versus FDD and Timing Advance 15

     1.11.1   Timing Advance and Power Back-off 16

1.12   LTE DL and UL Channels and Signals 16

1.13   UE Measurements 17

1.14   Signal Quality at High-Level 18

2 LTE UE HARDWARE AND INTERFACES 19

2.1   UE Modem Architecture 19

2.2   RF FEM 20

2.3   MIPI Interfaces Specific to LTE Hardware 21

2.4   Non-MIPI Interfaces Specific to LTE UE 22

2.5   UE Modem Architectural Split 23

2.6   RF FEM Losses 24

2.7   Noise Figure 25

2.8   RF and FEM Signal Quality 25

3 SIGNAL LEVELS, PAPR, AND CREST FACTOR 27

3.1   Waveforms Crest Factor and PAPR 27

3.2   OFDM Signals and PAPR 29

3.3   PAPR and CCDF for Noise 30

3.4   OFDM Signals in Downlink and Uplink 32

3.5   Relating Rx PAPR with AGC 33

3.6   Signal Level Examples 34

3.7   Noise Levels and Noise-Level Combining 37

4 LTE FREQUENCY AND CLOCK SOURCES 39

4.1   Clock Sources for LTE Operation 40

  4.1.1   Crystal Oscillator 40

  4.1.2   TCXOs 41

  4.1.3   OCXO 43

  4.1.4   GPS Derived Clock 43

4.2   Frequency Synthesizers 43

  4.2.1   Distinct Tx, Rx, and Clock Synthesizer Correction 44

4.3   Int-N and Frac-N Synthesizers 45

  4.3.1   Frac-N Synthesizers 45

  4.3.2   Frac-N Synthesizers Resolution for LTE Bands 46

4.4   Phase Noise and Jitter 47

  4.4.1   Phase Noise Frequency Regions 47

4.5   Phase Noise and Nonlinearity 48

4.6   Jitter Math from Phase Noise 49

4.7   Jitter Influence on Sampled Signals SNR 51

4.8   Power Dissipation, Voltage, and Temperature 51

4.9   Summary on LTE System Clock Generation 52

5 FREQUENCY ERRORS AND SIGNAL QUALITY 53

5.1   Doppler Shift in LTE System 54

5.2   Frequency Offset Estimation 56

5.3   Frequency Correction in the Signal Path 58

5.4   Carrier Frequency Offset Error and SNR Loss 58

5.5   Sampling Clock Offset (SCO) 60

5.6   Establishing Reference Quality with Frequency Errors 62

5.7   Conclusive Remarks 64

6 RECEIVER I-Q IMPAIRMENTS 65

6.1   I-Q Imbalance Illustrations 67

6.2   DC Offset Influencing the Signal Dynamic Range 69

6.3   Modeling the I-Q Imbalances 70

6.4   I-Q Imbalance Mathematical Formulation 72

6.5   I-Q Imbalance Numerical Examples 74

6.6   Trade-offs in Phase and Amplitude Balance 76

6.7   Time Delay or Clock-Skew Imbalance..77

6.8   Estimating Rx I-Q Imbalances in Frequency Domain..............80

6.9   Estimating Rx I-Q Imbalances in the Time Domain 81

6.10   Imbalance Correction 82

6.11   LTE BW and Band Edges 83

7   TRANSMITTER I-Q IMBALANCES, ESTIMATION AND CORRECTION 85

7.1   Amplitude Precomp Reduces the DAC Dynamic Range 87

7.2   I-Q Imbalance Correction by Design...87

7.3   Tx Imbalance Digital Correction Architecture 88

7.4   Time Delay Correction...89

7.5   Manual Characterization of I-Q Imbalances.............................89

7.6   I-Q Imbalances Estimation with BiST.......................................90

7.7   Tx LOFT and DC Offsets 91

7.8   Tx LOFT Estimation with Tx to Rx Loopback 91

7.9   Tx Amplitude and Gain Imbalance with Tx to Rx Loopback 93

     7.9.1   Tx Group Delay..94

8   TRANSMITTER GAIN MANAGEMENT AND TIME ALIGNMENT 95

8.1   Power Control with SRS in FDD and TDD 95

8.2   Tx Chain Gain and Loss Distribution 97

8.3   Timing Delay Variations in Tx 99

8.4   Tx Time Alignment Programming...100

8.5   Chapter Summary..101

9   RECEIVER AGC AND BLOCKERS 103

9.1   Rx Automatic Gain Control..103

9.2   Rx Time Alignment 105

9.3   Gain Control Processing 105

     9.3.1   Signal Strength Estimation 106

     9.3.2   Peak Signal Processing for AGC...............................106

     9.3.3   Signal Strength from Reference Tones 108

9.4   DC Offset Issues with AGC...108

9.5   Blockers 109

     9.5.1   Desensitivity..109

     9.5.2   Adjacent Channel Selectivity (ACS) 109

     9.5.3   In-Band Blockers 111

9.5.4   Out-of-Band (OOB) Blockers 113

9.5.5   Narrow-Band Blockers 114

9.5.6   Blockers and AGC 115

9.6   Chapter Summary 116

10 SIGNAL-QUALITY CHARACTERISTICS AND MEASUREMENTS 117

10.1   Transmitter Characteristics 117

10.2   Receiver Characteristics 120

10.3   EVM and SNR 121

10.4   Tx ON/OFF Mask and Transient Time 123

10.4.1   ACLR 123

10.5   Signal-Quality Connections Summary 124

10.6   Signal-Quality Measurements 124

10.6.1   LTE Testing with Call Generators 124

10.6.2   Low-Level Measurements 127

10.6.3   Typical Production Setup 127

10.6.4   Extended Lab Testing and Interop 128

10.6.5   Certification and Carrier Tests 129

10.7   Chapter Summary 130

11 RADIO LINK BUDGET FOR LTE SYSTEM 131

11.1   UE versus eNodeB in Terms of Radio Parameters 131

11.2   REFSENS for Uplink and Downlink 133

11.3   Received Power and Path Losses 134

11.4   UL Link Budget 136

11.5   Downlink Link Budget 138

11.6   Uplink Timing Advance 139

11.7   Conclusive Remarks 141

12 IOT SYSTEM AND SIGNAL-QUALITY 143

12.1   LTE-based IoT Categories 145

12.2   IoT Device Hardware Architectural Trade-offs 145

12.3   Power Dissipation and Peak Power Handling 146

12.3.1   Peak Tx Power Example 147

12.4   Power Amplifiers (PA) and Power Supplies 147

12.4.1   ACLR 148

12.5   IoT Device Power Consumption and Battery Life 148

12.6   Supercapacitors 149

12.6.1   Power Supply Overview 150

12.6.2   Capacitor Charging and Energy 150

12.6.3   Energy Lost in Charging Resistor 151

12.6.4   Supercap Ceff and ESR 152

12.6.5   Supercap Tolerance, ESR, and Temperature.....................152

12.6.6   Single versus Dual-Cell Supercaps....................................152

12.6.7   Energy Harvesting and Leakage 152

12.7   Narrow-Band LTE Signal Levels 153

12.8   Narrow-Band—Frequency and Clock Sources.......................153

12.8.1   Synthesizers 154

12.9   Narrow-Band—Slot and Frame Alignment............................154

12.10   Narrow-Band—Gain Controls..155

12.11   IoT Device Size and Antenna Size 155

12.11.1   Active Antenna and Tuning 156

12.12   IoT Device Testing 157

12.12.1   IoT Device Coexistence 158

12.12.2   IoT Device Environmental Conditions............................158

12.13   Signal-Quality Miscellaneous Connections 158

Bibliography...**160**

References by Standards 160

References by Author Index 161

References by URLs 162

Glossary...**171**

Acknowledgments

I enjoyed working at a microlevel on the long-term evolution (LTE) system, being one of the complex cellular/mobile/wireless systems. I was lucky to get an additional opportunity to work on improving the LTE modem and LTE-based Internet-of-Things (IoT) products' signal quality, which allowed me to improve my skills, expand my knowledge, and to contribute to multiple areas of the LTE system: physical-layer technology, mixed signals, analog, RF, RF FEM, radio links, coexistence with other wireless systems, overall product engineering, system integration, firmware, and testing.

I am grateful to the contributors/authors of LTE and mobile/wireless systems' technical literature and knowledgeable colleagues and project partners who allowed me to quickly learn and contribute to LTE and IoT technology in a better way.

Special appreciation goes to my daughter, Spandana Nagireddi, for proofreading and editing the book, and for researching on cover design, various publication and marketing options.

Several members reviewed the book material, and I thank Farbod Behbahani, Hardik Jain, Sattar Elyasi, Shengquan Hu, Sondur Lakshmipathi, and Sushil Gote for reviewing and sharing their technical views.

I sincerely thank Jonathan Siann, Marco Brambilla, and Hem Hingarh for their guidance and for posing tough technical problems in improving the LTE system and mobile IoT products' quality. At Sigma Designs Inc., I thank Thinh Tran, my project team, and project-execution partners for their encouragement and support during this publication.

I am indebted to my wife, Vijaya, for her persistent encouragement, accommodating my tight schedules, and taking several responsibilities such as editing figures and tables to make this publication happen, and to my son, Vamsi K. Nagireddi, for his contributions in book cover design, persistent follow-up and insisting me to complete this publication without allowing me to rest during weekends and holidays.

About the Author

Sivannarayana Nagireddi, PhD, is currently working as the Technical Director, Mobile IoT at Sigma Designs Inc., California. Dr. Sivannarayana and the associated project team developed LTE and other mobile/wireless technology derivative products for mobile modems and IoT devices.

Sivannarayana has been working on various core electrical-engineering technologies for the last twenty-five years. From 1986 to 1999, he worked as Research and Development (R&D) scientist—building complete systems for wireless communications, radars, and radar image processing. He researched and designed two air-borne radar systems, which enhanced the radar-system capability manifold. From 1999 to 2008, he worked as a VoIP architect and built complete SoCs (system on chips) and products for voice and VoIP that were widely deployed in Europe, Asia, and South America. From 2009 to 2012 he worked on various product derivatives of wireline, home networking, and gateway-interfaced femtocells, and from 2012 onward, his main focus area has been LTE and mobile/wireless systems, engineering multiple LTE and IoT systems, SoCs, and system integration for modem and IoT devices.

Sivannarayana graduated with a master's degree in electronics and communications engineering (ECE) from Osmania University, India. He was then awarded a PhD from the ECE department, Osmania University, with a focus on wavelet signal-processing algorithms and applications.

His favorite topics are mobile/wireless systems, communication systems, IoT, machine learning, end-to-end systems, signal chain analysis, as well as building systems/products and supporting them for successful use. He is a senior member of IEEE, a fellow of IETE, India, and reviewer for the journal Medical Engineering & Physics.

He has published about twenty journal and conference papers on multiple technical topics. He is the author of *VoIP Voice and Fax Signal Processing* (http://www.wiley.com/WileyCDA/WileyTitle/productCd-0470227362.html), a book published by Wiley, USA.

Preface

In the mobile world, long-term evolution (LTE) gained momentum and acceptance in the marketplace much quicker than the expected calendar dates, and the timing geared up very well with high-speed, low-power semiconductors that went into smart phones and multiple derivatives of LTE modem devices. Even though LTE is a complex mobile/cellular system with many advanced system- and signal-processing techniques put under one umbrella, the achieved (signal) quality of the LTE devices is much better and in many cases exceeds the 3GPP specifications by a big margin.

Nearly a decade of LTE effort is massively benefitting the world. Several valuable books have been published by experts in this field. While I was debugging several physical-layer and system-integration issues for identifying and improving the physical layer signal quality, I used to feel there should have been a good reference book on LTE signal quality with easy interpretations, some basic mathematics, and result tables. Several authors incorporated many selected chapters on the signal-quality aspects and signal impairments, and I thought I should give my part of easy-to-read contributions focused on this topic. The LTE subject is vast, and in this book, I have included selected topics on LTE and IoT systems, signal quality, analysis, basic mathematics and interpretations, illustrations, tabulated results, signal-quality relevance with 3GPP specifications, overall product architectural aspects, hardware, testing, and measurements for signal quality.

This book is organized into twelve chapters. Chapter 1 is an introduction to the LTE system. This is very basic material for continuity of the topic, and the reader may have to refer to additional LTE literature to get a deeper understanding of the overall LTE system.

In chapter 2, LTE user equipment (UE), architectures, and interfaces are presented, including the architectural evolution, multiple levels of integration, and various benefits. The mobile industry processor interface (MIPI) non-MIPI, and their various trade-offs and alternate options are illustrated.

In chapter 3, signal levels interpretations, peak-to-average-power ratio (PAPR), and characterizing the orthogonal frequency-division multiplexing (OFDM) through PAPR and crest factors are presented, including complementary cumulative distribution function (CCDF), signal-level combining, and noise-level combining.

In chapter 4, LTE operations carrier frequency, clock source requirements, various options, phase noise, jitter, int-N and frac-N synthesizers, and various quality-related aspects are presented. In continuity with chapter 4, chapter 5 addresses various clock source errors in oscillators, sampling clocks, and Doppler shifts and how to correct them to meet 3GPP specifications and to exceed the quality expectations.

Chapters 6 and 7 present I-Q imbalances, DC offsets, time offsets for the receiver (Rx), and the transmitter (Tx). A systematic, simpler arithmetic is included to show various options to estimate and cancel the I-Q gain and phase imbalances, DC offsets, and time offsets.

The key point for Tx is the gain control/management and timing alignment from baseband samples to RF power. Various issues and approaches to achieve better transmission quality are presented in chapter 8. A significant part of the chapter also includes the gain adjustments for various signals and ON/OFF issues in the Tx signal path. Chapter 9 mainly focuses on Rx gain control. The gain control algorithms use multiple hooks for the signal level, reference tones power, and PAPR at various thresholds. All these signal parameters are addressed with simple formulation and illustrations. The blockers, Tx leakages, and Rx dynamic range aspects are also significantly interpreted in this chapter.

Chapter 10 has several sections on multiple signal quality and measurement philosophy starting from low-level physical-layer measurements to production measurements. Examples are included with typical applicable test setup, and various capabilities are listed to look in the test setup.

Chapter 11 is an information chapter on various parameters involved with link (range coverage) budget. Several illustrations and easy-to-read plots and tables are included that interpret results up to the 100 km range under various propagation models. The relevant signal-quality sections are also part of this chapter.

Chapter 12 is fully dedicated to general characteristics and key architectural and implementation issues of Internet-of-Things (IoT). Many topics relevant to LTE IoT and generic IoT family implementations are included in this chapter.

I prepared this book with a blend of practical aspects, quality issues, system analysis, and basic math, with interpretation of the results in an easy way, including more than fifty figures and plots and more than twenty-five tabulated results with relevant interpretation. Overall I believe this book is an easy read and a quick reference of worked-out tables, and one should be able to read it through within a few hours. I hope this book will benefit a broad spectrum of engineers, from beginners to LTE and IoT test and system integration members, system engineers, hardware and firmware debugging teams, and product managers. The easy interpretation and examples will also help the non-physical-layer members in getting a good overview of what is looked upon on the physical-layer side to improve the signal quality.

I stayed with 3GPP release 9 for LTE material and newly evolving material on the IoT as LTE-M and NB-LTE. The IoT standards are still evolving; hence, only key aspects are listed in the book without going into all the specifics. The signal-quality issues and techniques are generally portable with extra dimensions of MIMO and LTE-advanced features. As mentioned previously, I have covered a few selected topics and hope to expand in future on a few more sections and also several extended sections/chapters specific to mobile IoT signal quality and analysis.

1 LTE SYSTEM

Long-term evolution (LTE) is a packet-switching-based mobile/cellular system known as 4G mobile service. The interfacing device, usually an LTE modem known as user equipment (UE), could be part of a smart phone, tablet, personal computer (PC), automobile, Internet-of-Things (IoT), or any other appliance. UE is a slave device to the LTE base station, which is E-UTRAN (evolved universal terrestrial radio access network) Node-B, also known as evolved Node-B, abbreviated as eNodeB or eNB. The LTE service provides Internet protocol (IP)–based connectivity between the UE and packet data network (PDN). The typical functional architectural representation of UE, eNB, and other supporting components of LTE deployment are shown in figure 1.1 [3GPP 36.401, URL (ALU1), URL (EPC-Arch)] and described as follows:

- UE is the user equipment that interacts with eNB using radio interfaces.
- eNB is the E-UTRAN that is the main radio access network (RAN) to the UE.
- EPC is the evolved packet core that interfaces with IP and IP multimedia services (IMS) networks.
- UE, eNB (E-UTRAN), and EPC are together referred to as the evolved packet system (EPS).
- The key components of EPC are S-GW, MME, P-GW, and PCRF.
  - S-GW: The UE IP packets are transferred to serving gateway, which provides local mobility when a UE moves across eNBs.
  - MME: The mobility management entity is the control node for signaling that performs bearer and connection management.
  - HSS: The home subscription server contains user subscription data and access restrictions. HSS is usually shown outside the EPC.
  - P-GW: The packet data network gateway is responsible for UE IP allocation and used along with PCRF that provides IP network access to the UE.
  - PCRF: The policy control and charging rules function is for policy

control and decision making that ensures user subscription policy
and options.

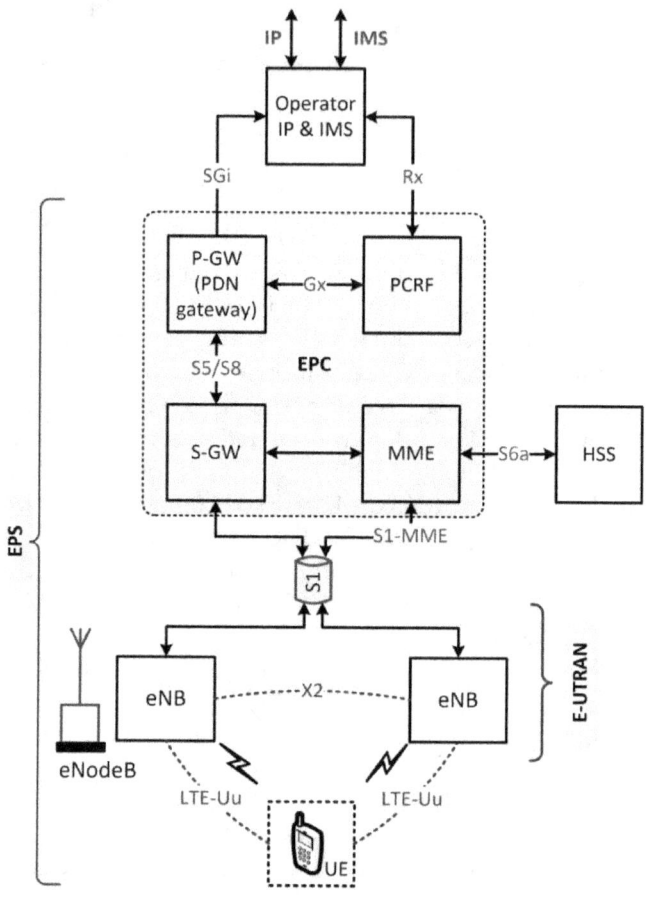

Figure 1.1. LTE system architecture—simplified functional representation.

The LTE architecture components are spread apart on a wide area network.
The interfaces follow the interface protocols [URL (EPC-Arch), URL
(RFwireless1)] marked at the high-level in figure 1.1. Many modules shown
in figure 1.1 may collapse into a high-speed processing server or can be
distributed across multiple servers or processing engines distributed across
a wide region.

1.1   LTE UE Architecture and Protocol Layers

The LTE layers from UE are represented in figure 1.2. The application layer sends data and control information through transmission control protocol (TCP) or user datagram protocol (UDP). The nonaccess stratum (NAS) [3GPP 24.301] is the control plane that supports signaling, mobility of the UE, and session management to maintain UE connectivity. The application data through the IP layer go through the access stratum (AS). The AS [3GPP 24.302] consists of RRC, PDCP, RLC, MAC, and the physical (phy) layer. As marked in figure 1.2, RRC works as a controller to many modules.

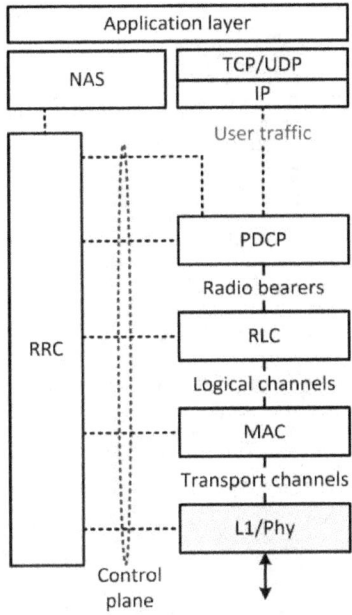

Figure 1.2. LTE UE protocol layers functional representation.

The packet data convergence protocol (PDCP) performs header compression, ciphering and deciphering of data, integrity protection, and transfer of data and maintenance of the sequence numbers [3GPP 36.323, URL (eventhelix1)].

The radio link control (RLC) layer performs transfer of RLC SDUs (service data unit) to upper layers; concatenation, segmentation, and reassembly of RLC SDUs, error correction through link layer automatic repeat request (ARQ), acknowledge mode for TCP and unacknowledged mode for UDP traffic, duplicate data detection, and so on as listed in [3GPP 36.322, URL (eventhelix2)].

Media access control (MAC) interfaces data on RLC logical channels and sends that to the physical layer on transport channels. MAC performs priority handling and scheduling, multiplexing, hybrid ARQ (HARQ) process, random access channel (RACH), feedback operation, timing advance and power control [3GPP 36.321, URL (eventhelix3)].

The physical (phy) layer interfaces data bytes with the MAC layer and sends radio waves [URL (nxp1), 3GPP 36.201, 3GPP 36.211, 3GPP 36.212, 3GPP 36.213, 3GPP 36.214].

In EPS architecture, UE interfaces with eNB and MME. NAS is a signaling layer that passes to MME through eNB. A functional representation of UE interfacing to eNB and MME is shown in figure 1.3.

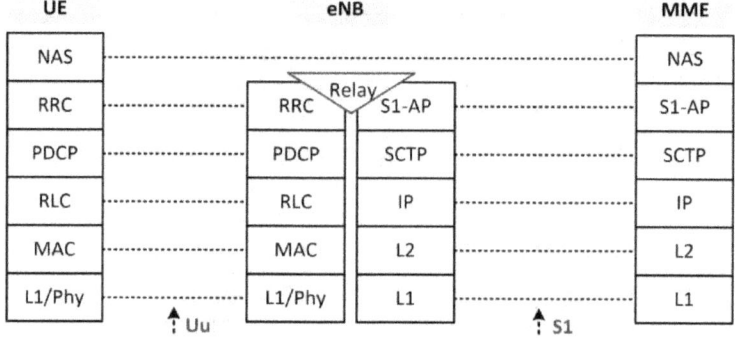

Figure 1.3. Representation of LTE system architecture layers.

Radio-access-network working groups (WG) have five main categories dealing with various specifications as described in references [URL (LTE-Wiley), URL (3GPP-RAN), URL (tech-invite)]:

1. RAN1 is layer 1—physical-layer specification.
2. RAN2 is layer 2 (MAC, RLC) and layer-3 specification.
3. RAN3 is overall UTRAN architecture.
4. RAN4 is radio performance and protocol aspects.
5. RAN5 is mobile terminal conformance testing.

1.2   LTE UE Uplink

LTE uplink (UL) sends data from UE to eNB. From applications and lower layers, streams of transport block data enter the physical layer. The top row of figure 1.4 denotes the channel coding, multiplexing, and interleaving. The data from transport layer are split into code words [Moray Rumney (2013)] and attached with cyclic redundancy check (CRC). On the basis of data size, the code-block segmentation splits data and attaches another CRC. When the data size is smaller, zero filling happens at the code-block segmentation. The data are channel coded for forward error correction using a turbo encoder. The coded data are rate matched to create required redundancy from the highest of 1/3rd to the least of 5/6th. The rate-matched code blocks are concatenated in the code-block concatenation module. The circular-buffer mechanism with the redundancy version (RV) is used to incrementally modify the redundancy and to facilitate the HARQ process. At this stage control data and measurements are multiplexed and created as code words.

The second row in figure 1.4 starts with scrambling, and the code words go through scrambling operation for additional error protection.

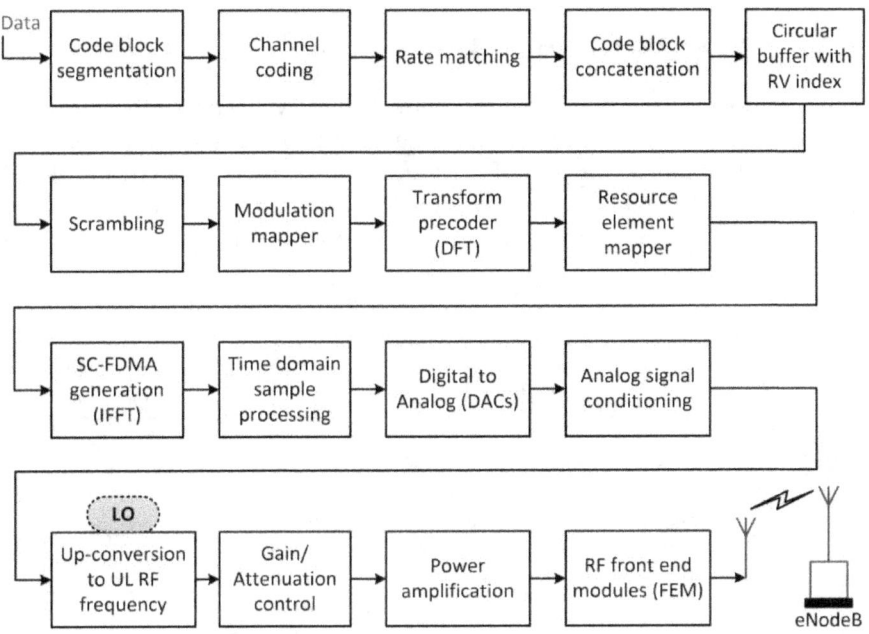

Figure 1.4. LTE UL physical layer.

The scrambled bits are modulation mapped, which means grouping the bits for quadrature phase shift keying (QPSK) and quadrature amplitude modulation (QAM) levels. As an example, 2 bits are grouped for QPSK, 4 bits for 16-QAM, and 6 bits for 64-QAM. The N points of mapper output are transform coded using discrete Fourier transform (DFT). The DFT output is properly spread into N points of inverse fast Fourier transform (IFFT) to generate time-domain samples. The chapters in this book deal with the operation from this IFFT stage. Time-domain samples in the format of in-phase (I) and quadrature phase (Q) are at the base sampling rate as decided by the LTE channel bandwidth (BW). The time-domain samples are processed for up-sampling, channel shaping filtering; gain control; precompensation for various impairments, such as DC offset, I-Q gain, and phase imbalances; and relative time delay between I-Q samples.

The digital I-Q samples are converted to analog usually at four times the base (IFFT block output) sampling, depending on the possible oversampling in hardware. The analog signal after required conditioning gets up-converted to LTE band carrier frequency. The radio-frequency (RF) signal goes through various stages of amplification/attenuation. The RF front-end modules (RF FEM) are used to deliver the required RF power levels from −50 dBm to +23 dBm at the antenna port.

1.3   LTE UE Downlink

The UE downlink (DL) is the reception from eNB. The RF signals enter the UE through antenna and go through RF FEM modules. The signals go through a low-noise amplifier (LNA) and get down-converted to analog I-Q channel baseband signals. The analog baseband conditions the I-Q signals and passes on to the analog-to-digital converters (ADCs).

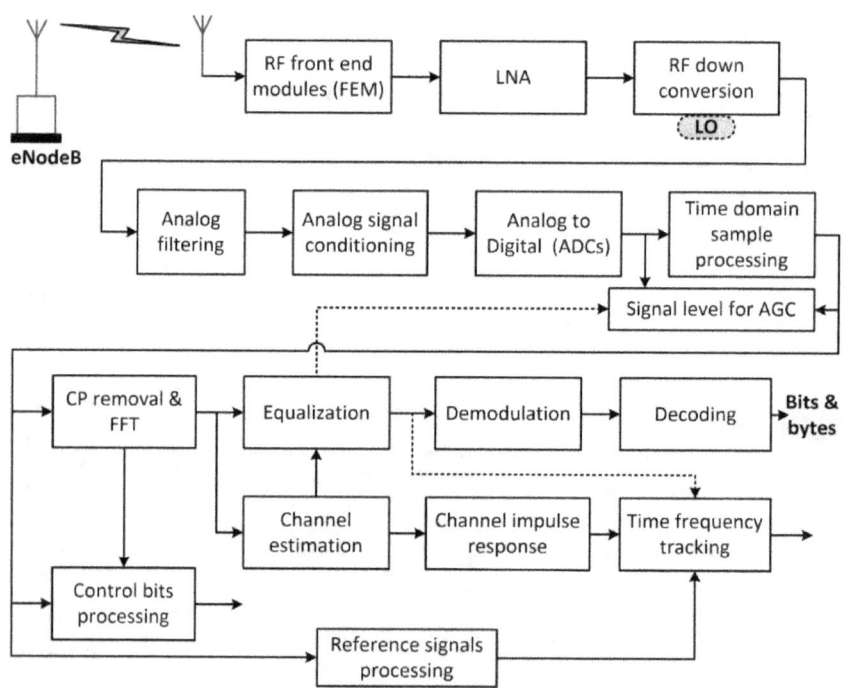

Figure 1.5. LTE DL physical layer.

ADCs usually oversample the signal by four times (depending on the oversampling capability of design) the base rate, and time-domain sample–based processing modifies the signal to the required base sampling rate. The digital samples are also enhanced for impairments, such as I-Q imbalances, DC offsets, time delay between I-Q samples, sampling rate alteration, frequency errors correction, and gain/attenuation as part of automatic gain control. At this stage several measurements are also made using I-Q samples. The main content of the book for DL sections deals with the operations up to this stage.

The time-domain samples go through multiple stages, the important operations being cyclic prefix removal, FFT operation, and reference signals processing. The frequency-domain data are used to estimate the channels, and OFDM tones are equalized with channel estimation data. In figure 1.5, the subsequent blocks shown with demodulation and decoding generate the data and control bits. Refer to [Chris Johnson (2012), Erik Dahlman, et al. (2011), Christina Gebner, Award Solutions] for several details on these topics.

1.4   Summary of Call-Flow Steps

The LTE call-flow summary is given here. Refer to [URL (Radisys), Award Solutions, Chris Johnson (2012)] for elaborated description. In an event of communication between eNB and UE, the following steps illustrate the call flow. UE turns ON, and on the basis of the preferred RF bands, UE keeps analyzing the RF signals with the center frequency raster (resolution) of 100 kHz (kilohertz). Once a particular RF signal with a few first-level required parameters matches through master information blocks (MIB) and system information blocks (SIB), the attach procedure starts. The attach starts with sending the RACH from UE and receives random access response (RAR) from eNB. With a few more steps, the basic communication gets established, referred to as radio resource control (RRC) connection [3GPP 36.331] establishment. The next steps to complete the connection establishment for data communication are NAS authentication and security, AS security activation, RRC connection reconfiguration, and UE capability transfer. After these steps, the data transfer may happen in UL and DL. When data are not present for a long time, RRC connection releases, and the UE stays in the idle state for a specified amount of time.

1.5   LTE Frame, Subframe, Slot, and Symbols

LTE operates on a basic frame (also known as LTE radio frame) of a time duration of 10 milliseconds (ms). Each frame is composed of 10 subframes

(SF)—each of 1 ms, and each subframe consists of two equal-duration slots. Each slot consists of 7 symbols in the case of normal cyclic prefix, and each symbol is of exactly 1/15,000 s, which equals to 66⅔ μs (microseconds), but for ease of representation, the symbol duration is represented in this book as 66.66 μs. The symbol tone spacing of 15,000 Hz (hertz) is reciprocal to symbol duration 66⅔ μs. The cyclic prefix (CP) with a normal-symbol subframe is 5.2 μs for the first symbol of a slot and 4.7 μs for the other symbols.

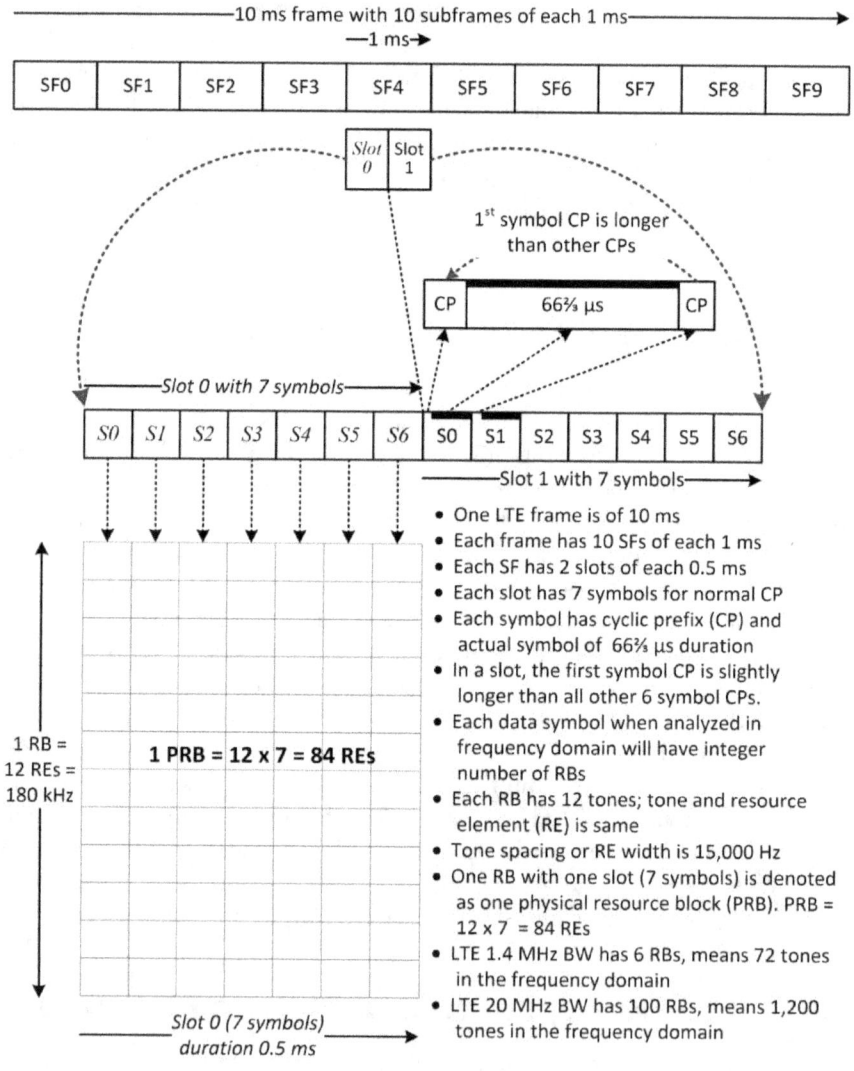

Figure 1.6. LTE frame structure.

With the extended cyclic prefix, the useful symbol duration is the same as 66.66 μs, but extended CP is 16.66 μs uniformly for all the symbols, and each slot accommodates 6 symbols (instead of 7 symbols in normal SF slots) with an equal CP-length (meaning time duration). Figure 1.6 shows the LTE frame structure and summary notes [Award Solutions, Chris Johnson (2012), Moray Rumney (2013)].

LTE UL (UE Tx) operates with a resolution of 1 subframe, and DL can work with a resolution shorter than 1 subframe—with symbol resolution. LTE UE can look at the reception in the first 1–3 symbols out of 14 symbols of a subframe, and if that DL subframe is not relevant for that UE, the UE can skip further processing of the received subframe signal, which is beneficial to IoT devices (chapter 12).

1.6   LTE BW and Time-Frequency Parameters

As listed in table 1.1, the LTE system operates at six different channel BWs of 1.4, 3, 5, 10, 15, and 20 MHz. Depending on the BW, the number of resource blocks (RBs) varies. Each RB has exactly 12 tones or resource elements (REs or tones) with tone spacing of 15,000 Hz (reciprocal of symbol duration 66.66 μs). The occupied (useful) BW is lower than the listed 1.4, 3, 5, 10, 15, and 20 MHz channel BW. Considering the example of 10 MHz LTE BW, the occupied BW is 9 MHz with 600 tones. The nearest 2^n point (where n is an integer) FFT or inverse FFT length for 600 tones is 1,024.

Table 1.1. LTE BW, RBs, tones, and interdependent basic time frequency parameters

Common parameters: Tones per RB = 12; Tone spacing: 15 kHz; Symbol duration: 66⅔ μs						
LTE channel BW MHz	Number of RBs	Max number of tones	Useful/ occupied BW MHz	Nearest 2^n FFT/IFFT points	Basic sampling frequency MHz	Sampling interval in ns
1.4	6	72	1.08	128	1.92	520.83
3	15	180	2.70	256	3.84	260.42
5	25	300	4.50	512	7.68	130.21
10	50	600	9.00	1,024	15.36	65.10
15	75	900	13.50	1,536	23.04	43.40
20	100	1,200	18.00	2,048	30.72	32.55

While processing all other tones outside 600 tones are discarded or used for possible indirect interpretation of noise and interference levels. To maintain 1,024 samples in a symbol duration of 66.66 μs, the sampling rate has to be 15.36 MHz, and the corresponding sampling interval is ~65.10 ns.

1.7   Adaptive Modulation and Coding (AMC)

Table 1.1 lists the maximum number of tones varying from 72 to 1,200 for LTE BW from 1.4 to 20 MHz. LTE operation may also use as low as 1 RB that makes it just 12 tones. The tones are QPSK, 16-QAM, and 64-QAM modulated depending on data and various quality parameters. QPSK uses 4 different constellation points to send 2 bits on each tone. The 16-QAM uses 16 constellation points to send 4 bits and 64-QAM for sending 6 bits per each tone. The modulation bits on QPSK, 16-QAM, and 64-QAM are coded to create redundancy to adapt to channel and signal conditions [Erik Dahlman, et al, (2011), Chris Johnson (2012), Moray Rumney (2013)]. For data bits the coding rates used are from 1/3 to 5/6, and the lower coding rate 1/3 provides less throughput (data transfer bits in a second) and works better for less quality signal. For control information the coding rates go as low as 1/16, and the coding rate for control bits is implemented as a combination of coding and duplicate information. The channel quality indicator (CQI) is closely related to the signal-to-noise (plus interference and distortions) ratio [URL (CQI)]. The modulation coding scheme (MCS) is decided on the basis of multiple parameters, such as CQI, ACK/NACK, data to be transferred, channel BW, and subscribed services by the user. Referring to ACK/NACK, the ACK is a positive acknowledgment, and NACK is a negative acknowledgment. As described in chapter 9 and 10, the throughput is measured with 95% of the maximum as reference. For AMC, the throughput thresholds will settle close to 90%.

What would happen if throughput is falling? The mechanism that quickly conveys the errors or throughput loss is ACK/NACK. And other key parameters are CQI and indirect channel quality information through sounding reference signal (SRS). In the DL from eNB to UE, the eNB may increase redundancy (by altering coding), and for maintaining the same throughput, eNB may use more RBs with less coding or allocate different location for resource blocks on the basis of the channel quality derived from SRS tones. SRS is the sounding reference signal sent from UE, and eNB accesses the channel path quality with frequency. SRS is generated on the basis of the UE parameters and initialized parameters from eNB. In the UL from UE to eNB, the loss of throughput or increased errors may result in increased UL transmission power, coding rate change, and spectral allocation change of RBs. Refer to [URL (lmk)] for an overview on channel coding and link adaptation, and chapter 11 for UL and DL link budgets.

1.8   Frequency Bands

The LTE system may use multiple frequency bands. A particular service provider may use a small part of a LTE band or may use two to three LTE bands depending on the region, spectrum allocation, and ownership. The LTE bands are listed in multiple references. A summary on bands is given in table 1.2 [3GPP 36.101, URL (free)].

The LTE UE is not the whole phone, and UE as a modem is used to interface with LTE tower (base station or eNB). The UE interfaces with the LTE base station—also referred to as eNodeB (or eNB). The signal that is transmitted from UE is the UL, and the signal received by the UE from eNodeB is the DL. Two main classifications of LTE mode of operation are given as time-division duplexing (TDD) and frequency-division duplexing (FDD). In TDD the same RF center frequency carrier is used for UL and DL, but UL and DL are time multiplexed.

1.9   Frequency-Division Duplexing

In FDD, a pair of RF carriers (RF center frequency) is used for UL and DL. Table 1.2 [3GPP 36.101, URL (free)] lists the major FDD carriers for UL and DL. The table contains DL and UL frequencies marked as low, middle, and high. The difference between high and low frequency is the total available BW in UL or DL. A service provider may get/use the whole band or a small part of the band. In FDD, UL and DL are positioned at different frequency location as both UL and DL operate simultaneously. The difference between the DL and UL frequencies is the duplex spacing, and the duplex spacing is required in FDD. The UE UL transmit power is much higher than the UE receive power. At UE, a small part of transmit power leaks (coupled with wired/conducted or wireless/radiated means) into its own receiver. The duplex frequency spacing helps improve the isolation between UL and DL. For most (but not all) FDD bands, DL is of higher frequency compared to UL. From the point of view of radio-link-coupling loss, the lower frequency will have less loss that benefits the UL from UE to eNB. As the spectrum usage is expanding worldwide, there are more spectral bands being approved for LTE deployments; hence, refer to the recent revision of 3GPP documents for new bands and their UL and DL frequencies.

Channel raster. LTE channel raster is 100 kHz, which means the center frequencies are integer multiples of 100 kHz [3GPP 36.521-1], decided by the eNB, and UE simply follows the eNB carrier.

Table 1.2. LTE FDD bands [3GPP 36.101, URL (free)]

Band	Downlink-DL frequency (MHz)			BW	Uplink-UL frequency (MHz)			Duplex spacing (MHz)
	Low	Middle	High	DL/UL (MHz)	Low	Middle	High	
1	2110	2140	2170	60	1920	1950	1980	190
2	1930	1960	1990	60	1850	1880	1910	80
3	1805	1842.5	1880	75	1710	1747.5	1785	95
4	2110	2132.5	2155	45	1710	1732.5	1755	400
5	869	881.5	894	25	824	836.5	849	45
6	875	880	885	10	830	835	840	45
7	2620	2655	2690	70	2500	2535	2570	120
8	925	942.5	960	35	880	897.5	915	45
9	1844.9	1862.4	1879.9	35	1749.9	1767.4	1784.9	95
10	2110	2140	2170	60	1710	1740	1770	400
11	1475.9	1485.9	1495.9	20	1427.9	1437.9	1447.9	48
12	729	737.5	746	17	699	707.5	716	30
13	746	751	756	10	777	782	787	−31
14	758	763	768	10	788	793	798	−30
17	734	740	746	12	704	710	716	30
18	860	867.5	875	15	815	822.5	830	45
19	875	882.5	890	15	830	837.5	845	45
20	791	806	821	30	832	847	862	−41
21	1495.9	1503.4	1510.9	15	1447.9	1455.4	1462.9	48
22	3510	3550	3590	80	3410	3450	3490	100
23	2180	2190	2200	20	2000	2010	2020	180
24	1525	1542	1559	34	1626.5	1644	1660.5	−101.5
25	1930	1962.5	1995	65	1850	1882.5	1915	80
26	859	876.5	894	35	814	831.5	849	45
27	852	860.5	869	17	807	815.5	824	45
28	758	780.5	803	45	703	725.5	748	55

1.10   Time-Division Duplexing

In TDD, a single carrier (center frequency) is used for both UL and DL. The subframes are time multiplexed between UL and DL with a time resolution of 1 symbol. TDD uses multiple configurations to share the time duration between UL and DL. A typical TDD frame structure is shown in figure 1.7 [Erik Dahlman, et al, (2011), Chris Johnson (2012), Moray Rumney (2013)]. In a 10-subframes frame, a few dedicated subframes are allocated to DL and UL. In a few selected subframes, known as special subframes (SSF marked as S), DL appears in selected symbols, known as DL pilot time slot (DwPTS), followed by time duration guard period (GP) and ends with UL pilot time slot (UpPTS). The guard period helps in distinguishing and allowing ramping up and timing advance (described in later chapters) of UL. The TDD maintains periodicity of 5 or 10 ms on the basis of configurations. Table 1.3 [3GPP 36.101, URL (free)] lists TDD bands that contain the same center frequency for UL and DL. The BW is the available frequency span between the low- and high-frequency edges of the band.

The TDD bands are listed in table 1.3. The TDD uses single carrier multiplexed in time between DL and UL with the resolution of 1 symbol. The BW marked in table 1.3 is for each LTE band, which is greater than or equal to LTE channel BW. LTE operates in this BW with a center frequency raster (resolution) of 100 kHz.

Table 1.3. LTE TDD bands

Band number	DL and UL frequency (MHz)			Bandwidth (MHz)
	Low	Middle	High	DL/UL
33	1900	1910	1920	20
34	2010	2017.5	2025	15
35	1850	1880	1910	60
36	1930	1960	1990	60
37	1910	1920	1930	20
38	2570	2595	2620	50
39	1880	1900	1920	40
40	2300	2350	2400	100
41	2496	2593	2690	194
42	3400	3500	3600	200
43	3600	3700	3800	200
44	703	753	803	100

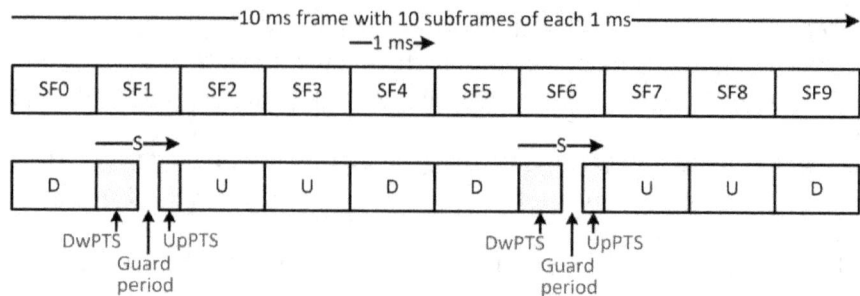

Figure 1.7. FDD frame (on top), and typical TDD frame (bottom).

Referring to table 1.4, TDD has multiple configurations [3GPP 36.211] to time multiplex between DL and UL. The subframe marked U is UL, D is DL, and S is special subframe used for both DL and UL with a guard period. If the DL data rate is higher, the configuration may use more D subframes and only few U subframes. 3GPP defined basic timing interval Ts as 32.55 ns [3GPP 36.211] (numerically equals to 1/30.72 MHz). As shown in table 1.1, the 30.72 MHz is the base sampling frequency for 20 MHz LTE channel BW. All the LTE timing intervals are calculated with Ts resolution even for LTE lower BW of 1.4 MHz.

Table 1.5 lists the SSF configuration that lists DwPTS, GP, UpPTS, and their duration in terms of Ts [3GPP 36.211, Moray Rumney (2013), Chris Johnson (2012)]. The timing advance span is part of the guard period of SSF. The farther the UE from eNodeB, the TDD SSF configuration with longer GP is selected. The relation between UE and eNB distance and the timing advance are described in chapter 11.

Table 1.4. LTE TDD subframe configuration [D:DL, U: UL, S: SSF subframes]

TDD UL-DL configuration	DL to UL switch point periodicity	Subframe number [0–9]									
		0	1	2	3	4	5	6	7	8	9
0	5 ms	D	S	U	U	U	D	S	U	U	U
1	5 ms	D	S	U	U	D	D	S	U	U	D
2	5 ms	D	S	U	D	D	D	S	U	D	D
3	10 ms	D	S	U	U	U	D	D	D	D	D
4	10 ms	D	S	U	U	D	D	D	D	D	D
5	10 ms	D	S	U	D	D	D	D	D	D	D
6	5 ms	D	S	U	U	U	D	S	U	U	D

Table 1.5. LTE TDD special subframe (SSF) configurations [*3GPP 36.211 release-11]

TDD SSF configuration	Number of Ts			Number of Symbols			
	DwPTS	GP	UpPTS	DwPTS	GP	UpPTS	Total in SSF
0	6,592	21,936	2,192	3	10	1	14
1	19,760	8,768	2,192	9	4	1	14
2	21,952	6,576	2,192	10	3	1	14
3	24,144	4,384	2,192	11	2	1	14
4	26,336	2,192	2,192	12	1	1	14
5	6,592	19,744	4,384	3	9	2	14
6	19,760	6,576	4,384	9	3	2	14
7	21,952	4,384	4,384	10	2	2	14
8	24,144	2,192	4,384	11	1	2	14
9*	13,168	13,168	4,384	6	6	2	14

1.11   TDD versus FDD and Timing Advance

In this section, the summary is given of timing advance to relate with TDD special subframe configurations. The eNodeB sends DL to multiple UEs with the same time alignment. With reference to simplified representation of UE1 and UE2, the DL is received at UE2 with 166.66 μs delay and at UE1 with 33.33 μs delay. The eNodeB expects UL from UE1, and UE2 to align at eNodeB. If both UE1 and UE2 send UL at the same time instance, the UE2 UL reaches eNodeB 166.66 − 33.33 = 133.33 μs later with reference to UE1, which is not acceptable. To avoid this time delay difference, UE2 time advances the UL transmissions. The UE2 is not knowledgeable about how much time advance to incorporate, and this information comes from eNodeB, and UE2 has to simply implement the communicated timing advance. In general every active UE gets timing advance updates that can be different from each other UE.

If UE1 and UE2 are operating on FDD, all the subframes are time advanced to the required value. In TDD, the UL and DL time slots are multiplexed in the same special subframe. For this reason, the multiple configuration in TDD special subframe helps to use up to the 100 km range. For long range, the timing advance is more, and TDD configuration uses SSF with longer guard period, managed by eNodeB (figure 1.8).

Notes: In the initial cell search, physical RACH (PRACH) uses multiple formats. PRACH format 0 is for the 14 km range, and PRACH format 3 allows the range of 100 km. The received signal strength decides the use of 100 km capability, even though PRACH format 3 allows from timing.

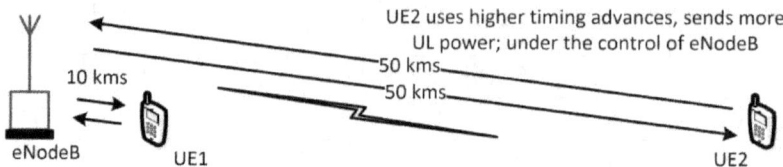

Figure 1.8. eNodeB serving UEs close in and far away.

1.11.1   Timing Advance and Power Back-off

As continuity of timing advance, the UE1 and UE2 have to maintain similar matching UL power reaching at eNB. Assuming UE1 and UE2 are transmitting the same power, the UE2 UL undergoes more coupling (path) loss compared to UE1 UL. The eNodeB keeps sending commands to UE2 to increase the UL power to (approximately) match UE1 and UE2 UL received (usually per RB) power at eNodeB receiver. The UE and eNodeB may not use timing advance to calculate power back-off, but they are correlated. The farther the UE2 from eNodeB, the higher the timing advance, and the UE2 is likely to receive messages for higher UL Tx power.

1.12   LTE DL and UL Channels and Signals

The DL (eNB to UE) channels and signals [Chris Johnson (2012), Christina Gebner, Moray Rumney (2013)] are summarized as follows:
- Physical synchronous signal (PSS) is for slot alignment of the LTE subframe, obtains part of cell identity (CID), and as explained in chapter 5, is also used for frequency errors (with reference to eNodeB) estimation.
- Secondary synchronous signal (SSS) is for frame synchronization and obtains part of cell ID information.
- Physical broadcast channel (PBCH) is for MIB.
- Physical downlink control channel (PDCCH) is for UL and DL resources information and transmit power control (TPC).
- Physical control format indicator channel (PCFICH) is for indicating PDCCH resources.
- Physical downlink shared channel (PDSCH) is for data and signaling traffic.

- Physical hybrid ARQ indicator channel (PHICH) is for conveying acknowledgments ACK/NACK for the UL.
- Physical multicast channel (PMCH) is used for multimedia broadcast and multicast services (MBMS).

The UL (UE to eNB) channels and signals [Chris Johnson (2012), Christina Gebner, Moray Rumney (2013)] are summarized as follows:

- Physical random access channel (PRACH) is for random access preambles.
- Physical uplink control channel (PUCCH) is for uplink control information (UCI).
- Physical uplink shared channel (PUSCH) is for data, control, and signaling messages.
- Sounding reference signal (SRS) is used to access channel quality and, if needed, to help frequency-dependent scheduling.
- Demodulation reference signal (DMRS) is used to establish demodulation of UL channels and applicable to both PUSCH and PUCCH.

Multiple extensions of the channels and signals may be used in LTE-A (advanced), LTE-B (beyond), LTE-M (machine to machine), and NB-LTE (narrow band LTE) specifications.

1.13   UE Measurements

UE will conduct measurements on the DL signal for internal consumption and also to interact with eNB. UE derives several parameters that are for internal use for better demodulation of DL signals or to send precompensated UL. UE measurements are used to optimize gain control, link adaptation, handover procedures, self-calibration, diagnostics, and so on.

The popular measurements derived from received signals [URL (IJWMN)] in the DL are reference signal received power (RSRP), received signal strength indicator (RSSI), reference signal received quality (RSRQ), signal-to-noise ratio (SNR), SNR with interference (SINR), and channel quality indicator (CQI). The SNR, SINR, and CQI use multiple proprietary techniques depending on how the baseband algorithms are implemented. RSRP is derived from the symbols that contain reference tones—also commonly known as pilot tones. The average pilot tone power is normalized using the receiver gain and calibrated gain errors. RSRP is relatively interference free. RSSI represents all the tones in the symbol and includes interference. RSRQ is derived from RSRP and RSSI.

1.14   Signal Quality at High-Level

Taking the example of the UE, quality is the perception and subjective things at the first level. The perception is created by previous history of the product family, ratings of the product, and current enhancements. At the next level, the quality aspects are the look and feel of the device, the user interface (UI), power dissipation (battery life) if applicable, and continuous operation without hanging, disconnects, and speed. At the next level, the quality concerns are from the number of bars (signal strength) in the coverage especially in the noncooperative environment, travel, and electromagnetic shadow regions.

The signal quality in this book stays much below the subjective nature of things. There are more objective things that go into digital samples, mixed signals, analog signals, RF signals, and air interfaces. The quality is decided by the parameters set by the 3GPP, test agencies, deployment carrier benchmarks, user feedback, and direct/indirect competition benchmarks. The objective quality indirectly helps improve subjective quality. Objective higher signal quality is achieved in multifold ways as summarized below:

- By architecture and design of hardware including system on chips (SoCs), and firmware.
- Fine-tuning the hardware, firmware, and overall system integration.
- Probes to look at temperatures, battery life, voltages, and various environmental, and electromagnetic conditions.
- Adaptive learning (not set by standards) to the environment, such as how the device is held, antennas, signal level, and adaptive tuning of the antennas.
- Learning and using the best out of learned conditions—a simplified machine learning—to overcome the repetition of bad performance.

2 LTE UE HARDWARE AND INTERFACES

Radio frequency (RF) front end (FE) is usually referred to as RF front-end module (FEM) or RF FEM. In the early stages of LTE UE modem development, the architecture was split into three main functions. The digital processing referred to as baseband (BB) or BB integrated circuit (BBIC), the analog and RF sections as RF integrated circuit (RFIC), and RF FEM. In this chapter, the relevance of many UE architectural blocks are included for proper interpretation of signal quality and losses.

2.1   UE Modem Architecture

As shown in figure 2.1, in the transmit (Tx) path, baseband digital section generates digital samples and interacts with mixed-signal digital-to-analog converters (DACs) that provide in-phase (I) and quadrature-phase (Q) analog signals. The analog signals are further enhanced/conditioned in analog baseband sections and up-converted to uplink (UL) frequency. The RF signal is further amplified to the higher power level using an RF power amplifier (PA), and PA is usually treated as part of the RF FEM.

In the receiver (Rx) path, the downlink (DL) RF signal from the antenna and RF FEM is amplified in low-noise amplifier (LNA) and down-converted to baseband analog signals. The analog baseband signals in I-Q path are further enhanced, and analog-to-digital converters (ADCs) convert the analog signals into I-Q digital samples. The baseband samples are processed in LTE baseband digital sections and higher layers processing. The baseband is associated with processors that control and program the RF and RF FEM.

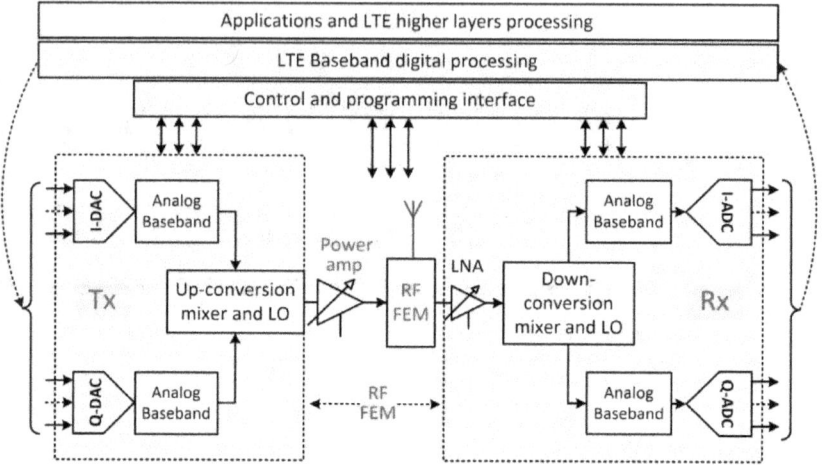

Figure 2.1. Representation of digital baseband, mixed signal, analog baseband, RF, and RF FEM.

The focus in this chapter is on RF FEM that interfaces with antenna and RF sections of Tx and Rx. Refer to [URL (Fujitsu), URL (limemicro1), URL (LMS7002M), URL (Berkeley)] for an example interfaces between RFIC and RF FEM. The RFIC is referred to as transceiver module, and the external discrete components are the RF FEM. The Tx section of RFIC provides signal to PA input, and Rx section receives signals at RFIC LNA input.

2.2 RF FEM

Figure 2.2 is a functional representation of LTE-FDD (frequency-division duplex) FEM with hardware blocks for two bands (shown as Band 13 and Band 7) with two Rx for main and diversity and one main path Tx. The RFIC Tx gives UL signal that goes through PA. The PA output goes to duplexer through matching network. In general inductor (L) and capacitor (C) based matching is applied at multiple stages in the FEM depending on the input and output characteristics of hardware components and printed circuit board (PCB) design. Architecturally, the duplexer has two passive filters, one for passing the Tx PA output to the antenna and the other for passing the receiver antenna output to the Rx. The Tx signal is strong that can leak into Rx path shown with dotted line from Tx to Rx at duplexer. Each band uses a separate duplexer as these are passive filters, typically used with one common antenna—assuming antenna works for wider frequency range. In the example diagram, Band-13 duplexer goes to antenna switch to use same antenna. When LTE system is using Band 7, the antenna is connected to Band-7 duplexer. In many implementations, directional coupler is used close to antenna. The coupler provides a tiny part of Tx power, and the envelope detector output is further conditioned (in analog or digital processing) to arrive at Tx power that will help in calibration, and closed-loop power control.

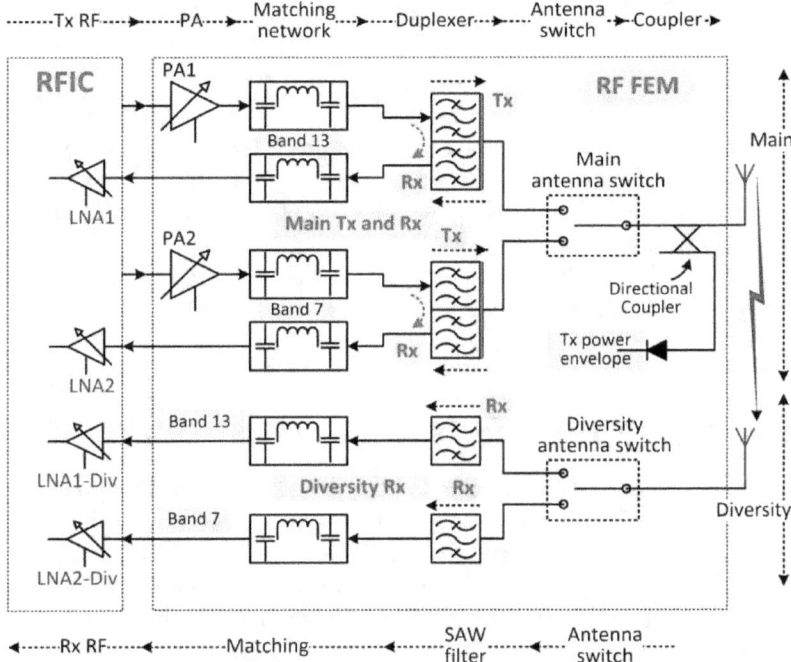

Figure 2.2. Functional representation of RF FEM for LTE FDD.

2.3   MIPI Interfaces Specific to LTE Hardware

The mobile industry processor interface (MIPI) arrived at mobile system intermodule hardware interfaces that minimize the connections between multiple discrete chips/modules of the mobile modem. With reference to [URL (MIPI-summary)], the interfaces relevant to LTE UE architectural modules are summarized as follows.

DigRFv4. Refer to [URL (MIPI-DigRFv4)] for digital RFv4 interface popularly used for multiple channels of Tx and Rx I-Q samples, control/configuration, and programming between BBIC and RFIC. In DigRFv4 inside BBIC is the master, and RFIC is the slave. The interface uses 7-lines and common ground. The example processor referenced at [URL (Fujitsu)] uses DigRFv4 interface to create communication between

BBIC and RFIC.

RFFE. The RF front-end (RFFE) interface uses three-pin serial with data, clock, and voltage input/output (VIO) [URL (MIPI-RFFE), URL (Fujitsu)]. The interface works at 26 MHz or lower rate. Using this RFFE interface, RFIC or any other modem control processor communicates with RF FEM devices, such as PAs, multiband multimode PAs [URL (triquint1)], and antenna switches.

LLI and HSI. The low latency interface (LLI) [URL (MIPI-LLI)] and high-speed synchronous serial interface (HSI) [URL (MIPI-HSI) are used to interface BBIC with application or higher layer processing engines.

Power supply interfaces. The system power management interface (SPMI) [URL (MIPI-SPMI)] is used to distribute power to multiple devices. The battery associated with power supply is supplemented with battery interface [URL (MIPI-Bat)]. The advanced power amplifiers use envelope-tracking (ET) power supply to optimize efficiency. The ET devices may use the eTrak interface [URL (MIPI-eTrak)]. Envelope-tracking power supplies are also built to interface with MIPI-RFFE [URL (ETPS-LM3290)] that works with either BBIC or RFIC.

2.4   Non-MIPI Interfaces Specific to LTE UE

Parallel I-Q interface. In the parallel I-Q interface, I and Q digital samples from BBIC and RFIC are interleaved on the same lines and sent from BBIC to RFIC in Tx UL path, and RFIC to BBIC in Rx DL path. MIPI-DigRFv4 was for I-Q samples, control, and programming; the parallel I-Q interface is used for I-Q digital samples. Refer to [URL (limemicro1), URL (LMS7002M), URL (ADI-AD9361)] for some examples on this interface. In general, devices support multiple options of I-Q samples interleaving to minimize the number of lines for samples transfer. For programming and control, SPI is the common interface.

JESD207. JESD207 is the JEDEC interface [URL (JESD207)] used to interface ADC and DAC data between RF and baseband chips. Commercial RFICs support multiple interfaces, such as parallel I-Q, and JESD207 [URL (MAX2580), URL (LMS7002M)].

Serial peripheral interface (SPI). Refer to [URL (limemicro1), URL (LMS7002M), URL (ADI-AD9361)] for the usage of SPI interface for programming the RFIC from BBIC. SPI works faster—commonly up to 50 MHz—for real time control of RFIC from BBIC.

General purpose input output (GPIOs). In the MIPI family of interfaces, RFFE is used to send control information from RFIC to RF FEM. As of year 2016, only few RF devices are supporting RFFE interface, and many devices are controlled through GPIO, generally referred to as IO controlled. GPIOs are used from BBIC or any other UE control processor.

2.5   UE Modem Architectural Split

As shown in figure 2.3, several popular architectural splits exist for distributing the hardware and processing functions, and the popular early stages of development architectural split being BBIC, RFIC, and RF FEM. As marked in figure 2.3, a particular product may use combination of MIPIs, non-MIPIs, and GPIOs to arrive at optimal solution.

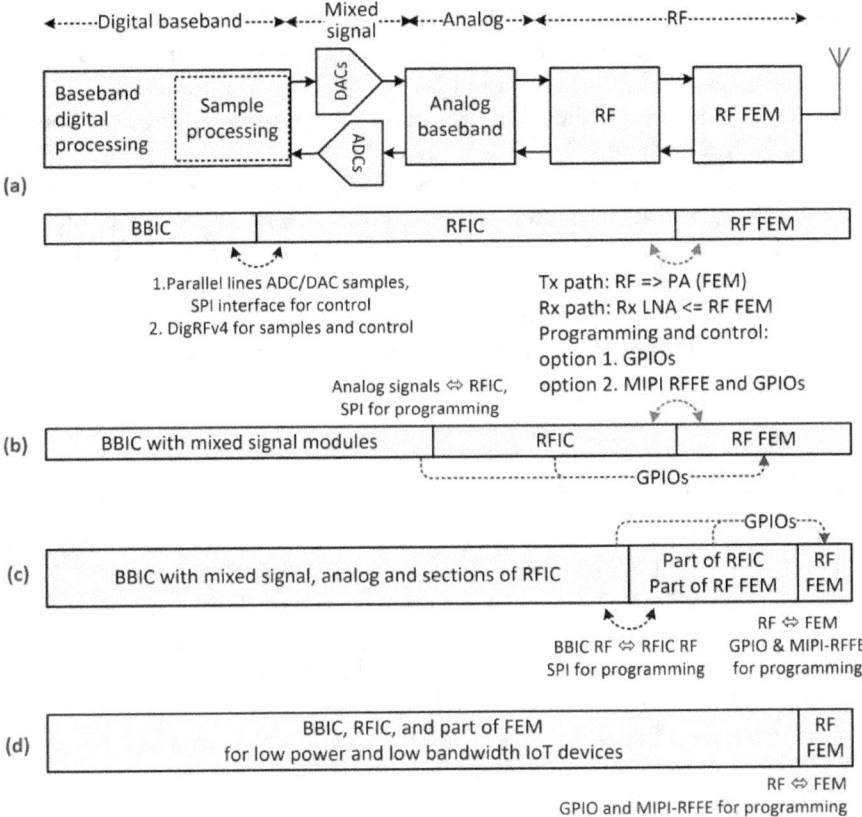

Figure 2.3. Various architectural splits of digital baseband, mixed signal, analog baseband, RF, and RF FEM. (a) Conventional architecture with BBIC, RFIC, and RFIC FEM split. (b) Mixed signal and parts of analog sections merged into the BBIC. (c) A part of RF functions moved inside the BBIC. (d) RF and part of FEM included as part of BBIC for IoT.

In figure 2.3(a) representation, BBIC sends digital samples to RFIC. The popular options to send samples are parallel digital bus usually multiplexes between I-Q samples or DigRFv4. DigRFv4 is an MIPI serial interface to transfer data and control/programming at a gigabit rate. When parallel I-Q data interface is used, a separate programming interface is required, and the popular option is the SPI or similar derivatives. Referring to figure 2.3(a), between RFIC and RF FEM, RF and analog signals are used. The programming and control of FEM may be through either GPIOs or combination of GPIOs and MIPI serial.

Figure 2.3(b) incorporates the mixed-signal sections into BBIC, and RFIC receives analog I-Q signal in place of I-Q digital samples. The benefit here is simplification of digital interfaces. In this approach, some sections of RFIC of figure 2.3(a) are incorporated into BBIC.

With the advancements in semiconductor technology, digital, mixed signal, analog, and RF are integrated into the same semiconductor chip, and figure 2.3(c) shows such composition. Architecturally, a part of RF FEM can also be part of RFIC.

The representation in figure 2.3(d) is for low-power devices such as IoT where many components can be collapsed into a single chip with reduced FEM.

2.6   RF FEM Losses

RF FEM creates losses to the signal. On the Tx side, the signal power is defined at the Tx antenna port. The losses and mismatches between PA and the antenna port can attenuate the Tx signal by usually 2–3 dB, depending on the PA output impedance characteristics, FEM components losses and mismatches, and the PCB structure. For example, consider FEM loss of 3 dB, and LTE UE maximum power of +23 dBm at the antenna port, the PA has to generate +26 dBm to account for FEM loss. The minimization of FEM loss is critical for the selection of PA, battery life, and to reduce UE power dissipation at PA.

Referring to figure 2.2, RF FEM loss in Rx is present between the Rx antenna port and LNA. The loss is usually accounted along with noise figure of the receiver. For an RFIC stand-alone noise figure (NF) of 3 dB (including LNA NF), and the FEM loss of another 2.5 dB makes the overall Rx NF as 5.5 dB. The RFIC and BBIC architectures as shown in figure 2.3, and BBIC algorithms will be optimized together to save even a small fraction of a dB. The optimization (reduction) of FEM Rx losses is one of the critical parameter in achieving the overall Rx performance.

2.7   Noise Figure

Noise figure in Rx with reference to table 11.1 is the signal-to-noise degradation from the antenna port to the receiver output, and this is majorly decided by the RF FEM and LNA. The LNA offers noise figure as degradation, and RF FEM from antenna connector offers multiple losses from—matching, baluns, switches, duplexers, RF filters, and PCB traces. The next level of noise figure contributions comes from RF and RF FEM nonlinearity that adds to the linear part. Noise and signal combining basic math is given in chapter 3. For overall product, the antenna efficiency and the product enclosure losses will also play a key role that becomes part of penetration loss as explained in chapter 11.

2.8   RF and FEM Signal Quality

The key signal-quality measurements associated with RF and RF FEM are illustrated in [URL (Aeroflex-7100), URL (R&S-1)]. With some extensions on top of [3GPP 36.521-1, 3GPP 36.101], the signal-quality items can be categorized as follows and expanded further in the next chapters:

- Transmission signal quality consists of signal level losses, imbalances, interference blockers, coexistence with other wireless/mobile systems, and nonlinearity-associated RF and analog sections.
- Timing of transmission, gain changes, blanking, control/programming differences through SPI, GPIO, and MIPI-RFFE.
- Frequency errors, frequency response, RF bandwidth, and signal leakage from Tx to Rx.
- Tx error vector magnitude (EVM) and out-of-band noise.
- Rx sensitivity and selectivity.
- Impedance match, tunable/adaptive FEM and antenna, and active antenna.
- Temperature and environmental effects.

3 SIGNAL LEVELS, PAPR, AND CREST FACTOR

Direct current (DC) or direct voltage has a steady/constant amplitude with time, and its root mean square (RMS or rms) voltage V_{rms} is the same as peak voltage (V_{pk}). The crest factor (CF) ratio is the peak voltage to the rms voltage V_{pk}/V_{rms} and in dB scale represented as $20 \log_{10}(V_{pk}/V_{rms})$. The crest factor is commonly used in signal analysis, and to represent amplitude growth in wireless/mobile orthogonal frequency-division multiplexing (OFDM) systems.

The crest factor is for voltage ratio, and the peak-to-average-power ratio (PAPR) is the power ratio. PAPR = (Peak power)/(Average power), and represented in dB scale as $10 \log_{10}$(Peak power/Average power). PAPR is always the ratio of the same power units, such as watts, milliwatts, or microwatts. The crest factor is related to PAPR ratio as

$$\text{PAPR} = \text{CF}^2, \quad \text{CF} = \sqrt{\text{PAPR}}, \text{ and } 20 \log_{10}(\text{CF}) = 10 \log_{10}(\text{PAPR}). \quad (3.1)$$

In wireless engineers' terminology, PAPR is liberally used in place of the crest factor and in general this is acceptable when represented in dB scale, but both are numerically not equal as a ratio. The average power in PAPR is calculated from rms voltage. For DC voltage the crest factor is 1, or $20 \log10$(voltage ratio) = 0 dB. In the case of power-based calculations, $10 \log10$[(Peak DC power)/(Average DC power)] = 0 dB.

3.1   Waveforms Crest Factor and PAPR

The sine wave has one-sided peak amplitude of V_{pk}, and its rms level is $V_{pk}/\sqrt{2}$ or $0.7071\ V_{pk}$; CF is $20 \log_{10}[V_{pk}/(0.7071\ V_{pk})] = 3.01$ dB, usually taken as 3 dB as close approximation. From electrical-engineering math, the average value of the sine wave is $(2/\pi)\ V_{pk} = 0.637\ V_{pk}$, which is not the same as the rms value used in calculating the crest factor. Refer to [URL (rfcafe)] for easy interpretation and formulation on rms and average voltage of sine wave. The square wave crest factor is unity similar to DC. The peak voltage and rms voltage of square wave are the same, and the PAPR ratio is unity (1) or 0 dB (figure 3.1).

Considering combination of two sine waves/tones f1, f2 of amplitude V_{pk} each, the combined signal PAPR is 6 dB, as listed in table 3.1. For combination of multiple tones with V_{rms1}, V_{rms2}, V_{rms3}, ..., V_{rmsn} the combined V_{rms} is sqrt($V_{rms1}^2 + V_{rms2}^2 + V_{rms3}^2 + \cdots + V_{rmsn}^2$).

Table 3.1. Signal voltages and PAPR

Signal waveform	Peak voltage	RMS voltage	Possible peak voltage	PAPR (voltage ratio)	Peak power 1-ohm	Average power in 1-ohm	PAPR (power ratio)	PAPR dB
DC voltage	V_{DC}	V_{DC}	V_{DC}	1	V_{DC}^2	V_{DC}^2	1	0
Square wave	V_{pk}	V_{pk}	V_{pk}	1	V_{pk}^2	V_{pk}^2	1	0
Sine wave	V_{pk}	$V_{pk}/\sqrt{2}$	V_{pk}	$\sqrt{2}$	V_{pk}^2	$V_{pk}^2/2$	2	3
Two sine tones at f1, f2	V_{pk1}, V_{pk2}	sqrt($V_{pk1}^2/2$ $+ V_{pk2}^2/2$)	$V_{pk1}+$ V_{pk2}	2 ($V_{pk} =$ $V_{pk1} = V_{pk2}$)	$4V_{pk}^2$	V_{pk}^2	4	6

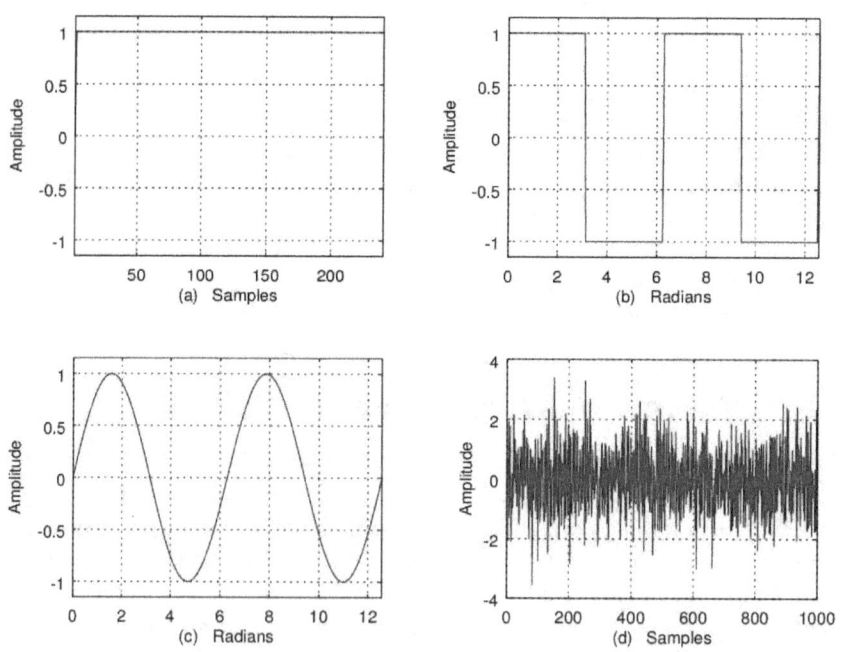

Figure 3.1. Signal waveforms. (a) DC. (b) Square wave. (c) Sine wave. (d) Random noise.

3.2   OFDM Signals and PAPR

For multitone OFDM (orthogonal frequency division multiplexing) signals, the highest peak occurs when amplitude from several tones build up with the same polarity at that particular time instant. When N-tone quadrature amplitude modulation (QAM) signals are combined in time domain, the time-domain I and Q signals are combined as per the modulation bit pattern incorporated on the tones. As shown in figure 3.2, 16-QAM signal will have three different amplitude levels and multiple phase components. Assuming the tones are uniformly distributed, the signal probability of getting highest peak from this 16-QAM signal is very low.

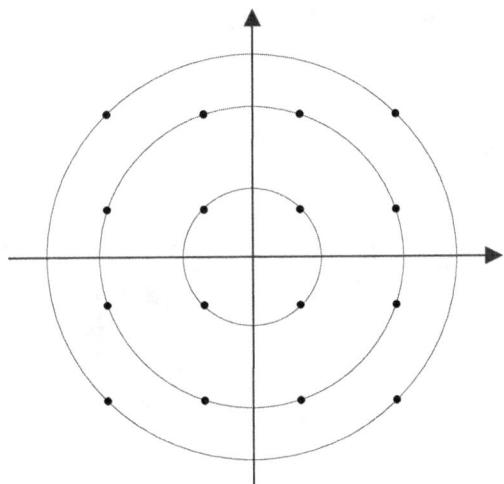

Figure 3.2. 16-QAM constellation representations (center 4, middle 8, and outer 4 points).

The I-Q samples from the OFDM modulated signal are approximated as Gaussian noise waveform [URL (nit), URL (noise)], and a small segment of simulated noise samples are plotted in figure 3.3(b). The probability density function (PDF or pdf) in figure 3.3(a) is the histogram of the samples shown in figure 3.3(b). The histogram in figure 3.3(a) is shown tilted to match the y axis as amplitude and x axis as the probability. Even though this is the true representation, the usual way is to see probability on the y axis as shown in figure 3.3(c). The OFDM signal or noise can have huge peak with very low probability, and the peaks may not be visible on a small samples size. In practical systems, a small percentage of peaks are clipped or modified to contain the PAPR to reasonable limits. The percentage of samples contained within different standard deviation (STD) values are listed here:

$1\sigma \Rightarrow 68.27\%$ is STD (σ) or rms value, and samples outside 1σ is 31.73%.

$2\sigma \Rightarrow 68.27 + 27.18 = 95.45\%$, and samples outside 2σ is 4.55%.

$3\sigma \Rightarrow 68.27 + 27.18 + 4.28 = 99.73\%$, and samples outside 3σ is 0.27%.

$4\sigma \Rightarrow 68.2 + 27.2 + 4.28 + 0.2636 = 99.9936\%$, and samples outside 4σ is 0.0064%.

3.3   PAPR and CCDF for Noise

The PAPR responses are shown as complementary cumulative distribution function (CCDF) plots. A good representation on the CCDF is given in [URL (Agilent-CCDF)].

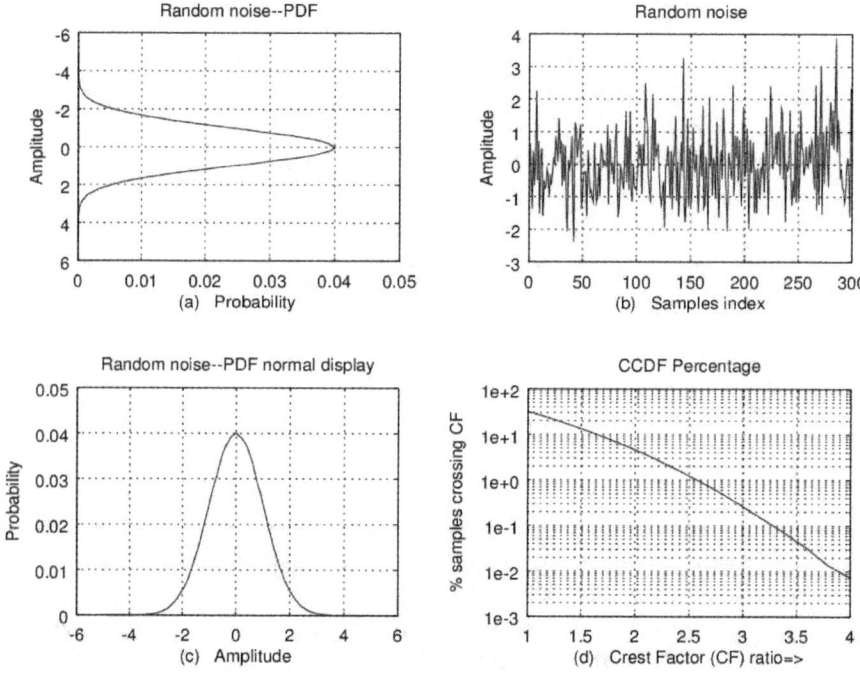

Figure 3.3. (a) Probability functions to match noise. (b) Gaussian noise. (c) Probability function—usual way of representation. (d) CCDF versus crest factor ratio.

Notes: To convey CCDF in a simplified way, CCDF is represented as inverse operation and mathematically as 1 minus CDF (1 − CDF), and cumulative distribution function (CDF) is the integration or summation of PDF. This leads to errors in the estimation of CCDF. The PAPR at 0 dB (1σ point is sample amplitudes from −1σ to +1σ), but as per CDF formulation, CDF for 1σ uses x-limits from −∞ to +1σ. Hence, CCDF as (1 − CDF) be erroneous. Referring to figure 3.4(a), the CCDF 1σ (PAPR 0 dB) is the points left side of −1σ and points right side from 1σ, and the CCDF 2σ (PAPR 6 dB) is the points left side of −2σ and points to the right side from 2σ.

CCDF shown in figure 3.3(d) is usually shown with the x axis in dB scale, which is easy to relate the measurements. The similar representation of figure 3.3(d) is shown in figure 3.4(b) with dB scales marked on the x axis. From figures 3.3(d) and 3.4(b), the applicable relations are

PAPR = CF^2, CF = $\sqrt{PAPR}$, and 20 $\log_{10}$(CF) = 10 $\log_{10}$(PAPR).

For understanding CCDF, figure 3.4(a) is shown with Gaussian probability function with zero mean. The zero mean and nonzero mean functions are represented as

$$y = \frac{1}{\sigma\sqrt{2\pi}}\, e^{-\frac{x^2}{2\sigma^2}} \quad \text{(zero mean)} \qquad (3.2)$$

$$y = \frac{1}{\sigma\sqrt{2\pi}}\, e^{-\frac{(x-\bar{x})^2}{2\sigma^2}} \quad \text{(nonzero mean)} \qquad (3.3)$$

where σ is the STD, σ^2 is the variance, and $\bar{x}$ is the sample mean. In figure 3.4(a), the x axis is the amplitude (when related with electrical signals), and the y axis is the occurrences of amplitude. Relating figures 3.3(c) and 3.4(a), 1σ (one sigma) is the standard deviation of the Gaussian PDF that represents the rms value of Gaussian noise.

As stated in the previous sections, CCDF is required for various values of PAPR. Figure 3.4(b) is CCDF versus PAPR in dB. In relating figures 3.3(d) and 3.4(b), both represent the same, except figure 3.3(d) is from noise samples, and figure 3.4(b) is from mathematical function given in equation (3.2). In the CCDF plots of figure 3.4(b),

1σ ⇒ 68.27% rms value ⇒ CF = 1, and PAPR = 1 or 0 dB is the starting reference that indicates on y axis 31.73% of the samples are outside the 1σ. 2σ ⇒ 95.45% ⇒ 4.55% are outside 2σ; CF = 2, and PAPR = 4 or 6 dB.

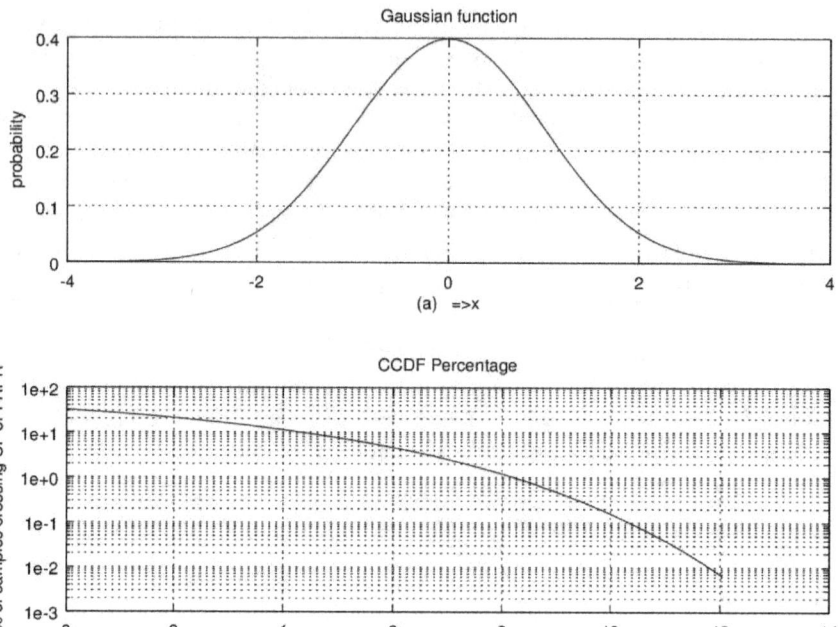

Figure 3.4. (a) Gaussian probability function. (b) CCDF versus PAPR in dB. From the plots the starting x axis 0 dB is 1σ point, the 6 dB is 2σ point, and 4σ is 12 dB point. The points exceeding 1σ is $100 - 68.27\% = 31.73\%$, *the starting point on the y axis at PAPR = 0 dB.*

3.4   OFDM Signals in Downlink and Uplink

In the uplink (UL), the OFDM signals are generated on the user equipment (UE). To reduce RF hardware power dissipation, it is important to reduce PAPR in the UE transmitter. In UL, the single carrier OFDM (SC-OFDM) is used for lowering PAPR. The SC-OFDM goes through the mathematical transformation of discrete Fourier transform (DFT) and then inverse fast Fourier transform (IFFT). With SC-OFDM in UL, the PAPR is slightly reduced compared to direct OFDM-based downlink (DL) from eNodeB.

The DL is a regular OFDM with higher PAPR. The UE receives signal with higher PAPR. Typical LTE signals PAPR numbers on DL time-domain I-Q samples are in the approximate range of 9–12 dB, and UL PAPR is in the range of 7–9 dB. The PAPR slightly increases with increased QAM levels among QPSK, 16-QAM, and 64-QAM; the number of tones in the symbol; and the QAM data phases present in that symbol. The RF signal is a combination of I-Q signals in the up-conversion operation, and

RF OFDM signal PAPR is lower than the baseband I-Q signal PAPR due to I-Q signal combining. The sine wave has PAPR of 3 dB. An I-Q OFDM signal stated with PAPR of 9 dB has 9 − 3 = 6 dB, more peak amplitude than single sine wave tone.

To measure the STD of noise or OFDM signal, often simple approximation can be used that closely matches the results within a fraction of dB error. The error is insignificant compared to signal dynamic range and automatic gain control (AGC) span of receiver:

$$s = \sqrt{\frac{\pi}{2}} \frac{1}{N} \sum_{n=0}^{N-1} |x_n|. \qquad (3.4)$$

The sum of the absolute value of I or Q samples denoted as x_n is multiplied with a constant $\frac{1}{N}\sqrt{\frac{\pi}{2}}$. The simplification reduces complexity to implement the signal strength estimation in hardware or on a processor.

3.5   Relating Rx PAPR with AGC

As explained in chapter 9 that signal level is controlled through gain variations. The first-level control is the rms signal level that corresponds to 1σ in the Gaussian PDF and CF factor of 1 or PAPR of 0 dB on CCDF plot. On the basis of the signal characteristics, such as the number of OFDM tones, QAM levels, and block of samples used for signal level estimation, the CCDF versus PAPR levels vary. To account for PAPR variations, a higher level of 10–12 dB headroom is created on top of rms level. To optimize PAPR headroom (means to provision lower headroom), it is essential to use PAPR measurements. The PAPR measurement approximations are described with the following example.

In this section, analog to digital converter (ADC) input analog signal is interpreted from ADC output digital samples. Assuming ADC output is in 12-bit numbers, to quickly access PAPR level at the ADC analog input point, the rms level of the ADC output samples is estimated. The PAPR can be estimated with a threshold detector. In threshold detector, a threshold of, say, 90% of the ADC maximum amplitude (ADCmax) is used that counts the samples exceeding 90% of the full scale. For a 12-bit twos complement ADC, the ADCmax is ~2^{11}. The PAPR crossing counter, counts/increments if abs$\{x(n)\}$ >0.9 ADCmax. (Note: The threshold 0.9 is 20 $\log_{10}(1.0/0.9) = 0.92$ dB away from full scale.) And this count is normalized with the number of samples used. This forms a point on the CCDF curve. If the number of PAPR count points is more than ~0.2%, then gain control headroom can be increased, and if the number of points is less than 0.2%, the headroom can be decreased. The headroom will translate to gain change as part of AGC, and the increase of headroom means reduction of the gain. The 0.2% (0.27% corresponds to $3\sigma \Rightarrow 9.54$ dB PAPR for Gaussian PDF) value used as an example for good signal quality. Often the headroom is intentionally increased to account for various margins, such as channel fading and gain errors. In the example above, the threshold has to be lower than 0.9ADCmax by the required additional margin.

3.6   Signal Level Examples

The basic unit of power is watt (W). LTE base stations, UEs, and small cells transmit RF power. Table 3.2 lists various power levels and their easy lookup in various convenient power units. In the table all the power levels are finally represented in dBm units.

The math is simple with the following relations: $1\text{ W} = 1,000\text{ mW} = 1\times10^6\text{ }\mu\text{W} = 1\times10^9\text{ nW} = 1\times10^{12}\text{ pW} = 1\times10^{15}\text{ fW}$, where mW is milliwatts, μW is microwatts, nW is nanowatts, pW is picowatts, and fW is femtowatts. Table 3.2 lists the corresponding dBm units. For the power P in mW, the power in dBm units is expressed as $10\log_{10}(P\text{ mW}/1\text{ mW})$:
$1\text{ W} = 1,000\text{ mW} = 10\log_{10}(1,000\text{ mW}/1\text{ mW}) = 30\text{ dBm}$
$1\text{ mW} = 0\text{ dBm}$, $1\text{ }\mu\text{W} = -30\text{ dBm}$, $1\text{ nW} = -60\text{ dBm}$, $1\text{ pW} = -90\text{ dBm}$, $1\text{ fW} = -120\text{ dBm}$.

To combine multiple signal or noise power levels, all the components have to be expressed in the same units. The power $P1, P2, P3, \ldots, Pn$ can be combined as $P1+P2+P3+\cdots+Pn$. When power levels are given in dBm units or any other dBx units, the power has to be converted back to non-dB units for combining.

Table 3.2. Power levels and dBm mapping

Watts	milli watt (P mW)	micro watts	nano watts	pico watts	femto watts	atto watts	10 log$_{10}$ (P mW) dBm	Remarks
40	40,000						46	LTE base station Tx
10	10,000						40	LTE base station Tx
1	1,000						30	Small and femto cells Tx power
0.2	200						23	LTE UE Tx max
0.1	100						20	
0.01	10						10	
0.001	1	1,000					0	
	0.1	100					−10	
	0.01	10					−20	
	0.001	1	1,000				−30	
	0.1×10^{-3}	0.1	100				−40	LTE UE Tx min
	10×10^{-6}	0.01	10				−50	LTE UE Tx OFF
	1×10^{-6}	0.001	1	1,000			−60	
	0.1×10^{-6}		0.1	100			−70	
	10×10^{-9}		0.01	10			−80	
	1×10^{-9}		0.001	1	1,000		−90	⇒ 1 pico watt
	0.1×10^{-9}			0.1	100		−100	LTE UE 1.4 MHz REFSENS ~0.1 pW
	10×10^{-12}			0.01	10		−110	
	1×10^{-12}			0.001	1	1,000	−120	⇒ 1 femto watt
	0.1×10^{-12}				0.1	100	−130	
	10×10^{-15}					10	−140	
	1×10^{-15}					1	−150	⇒ 1 atto watt
	0.1×10^{-15}					0.1	−160	
	10×10^{-18}					0.01	−170	⇒ 0.01 atto watt
	4×10^{-18}					0.004	−174	Noise floor k_BT BW dBm/Hz

Table 3.3 lists the combined power levels for the same dBx units or dBm units. In RF signal level calculations, dBm levels combining are more common. For combining $x1$ dBm and $x2$ dBm, the combined x dBm is given as $10 \log_{10}[10^{\wedge}(x1/10) + 10^{\wedge}(x2/10)]$. Basically dB units are converted back to power levels, and after summation the log scale is applied again. To combine k signals power level from $x1, x2, \ldots, xk$ in the same dB power units, use $10 \log_{10}[10^{\wedge}(x1/10) + 10^{\wedge}(x2/10) + \cdots + 10^{\wedge}(xk/10)]$.

Table 3.3. Signal level combining

x1 dBm	x2 dBm	Combined x dBm	dBs change |x-max(x1,x2)|	Remarks
0	−12	0.27	0.27	12 dB relative difference changes ~0.25 dB
0	−11	0.33	0.33	
0	−10	0.41	0.41	
0	−9	0.51	0.51	9 dB relative difference increases by ~0.5 dB
0	−8	0.64	0.64	
0	−7	0.79	0.79	
0	−6	0.97	0.97	6 dB relative difference increases by ~1 dB
0	−5	1.19	1.19	
0	−4	1.46	1.46	
0	−3	1.76	1.76	3 dB relative difference increases by 1.76 dB
0	−2	2.12	2.12	
0	−1	2.54	2.54	
0	0	3.01	3.01	both of equal level increases by ~3 dB
0	1	3.54	2.54	
0	2	4.12	2.12	
0	3	4.76	1.76	3 dB relative difference increases by 1.76 dB
0	4	5.46	1.46	
0	5	6.19	1.19	
0	6	6.97	0.97	6 dB relative difference increases by ~1 dB
0	7	7.79	0.79	
0	8	8.64	0.64	
0	9	9.51	0.51	9 dB relative difference increases by ~0.5 dB
0	10	10.41	0.41	
0	11	11.33	0.33	
0	12	12.27	0.27	12 dB relative difference changes ~0.25 dB

3.7   Noise Levels and Noise-Level Combining

The noise levels overview is given in [URL (noise)]. The LTE system uses multiple BWs starting from one tone of 3.75 kHz narrow-band LTE operation to 20 MHz for regular LTE. The noise levels for multiple noise BWs relevant to LTE are listed in table 3.4. The $k_B T$ per 1 Hz is −174 dBm/Hz where temperature (T) equals to 290 kelvin and k_B is Boltzmann constant [URL (noisecom)]. For other BWs, the noise power is given as −174 + 10 log$_{10}$ (BW in hertz) dBm. The noise levels $n1$, $n2$, $n3$ in dBm or dBc (dBc is the dB scale with reference to the carrier, also denoted as dBm/Hz) can be combined similarly to multiple signals combining. To combine k power levels from $n1$, $n2$, …, nk in the same dB power units, use 10 log$_{10}$[10^($n1$/10) + 10^($n2$/10) + ⋯ + 10^(nk/10)].

Table 3.4. Noise levels

$k_B T$ (1 Hz) dBm/Hz	Bandwidth (BW)	Noise power in dBm	LTE channel BW designation
−174	1 Hz	−174.0	Reference as dBc or dBm/Hz at room temperature
−174	1 kHz	−144.0	
−174	3.75 kHz	−138.3	Narrow-band LTE 1-tone
−174	7.5 kHz	−135.3	½ of 15 kHz
−174	15 kHz	−132.2	LTE 1-tone
−174	180 kHz	−121.5	LTE 1-RB (12 tones × 15 kHz)
−174	1 MHz	−114.0	BW = 1 MHz for reference math
−174	1.08 MHz	−113.7	Occupied BW of LTE 1.4 MHz
−174	1.40 MHz	−112.5	LTE 1.4 MHz
−174	2.7 MHz	−109.7	Occupied BW of LTE 3 MHz
−174	3.0 MHz	−109.2	LTE 3 MHz
−174	4.5 MHz	−107.5	Occupied BW for LTE 5 MHz
−174	5.0 MHz	−107.0	LTE 5 MHz
−174	9.0 MHz	−104.5	Occupied BW for LTE 10 MHz
−174	10 MHz	−104.0	LTE 10 MHz
−174	13.5 MHz	−102.7	Occupied BW for 15 MHz
−174	15 MHz	−102.2	LTE 15 MHz
−174	18 MHz	−101.4	Occupied BW for LTE 20 MHz
−174	20 MHz	−101.0	(*)LTE BW 20 MHz

> Notes: (*) In chapter-11, table 11.2 UE sensitivity $P_{REFSENS}$ for 20 MHz is listed as −94 dBm for Band 4 operation. Considering close to 5.5 dB noise figure of the receiver that includes receiver (LNA with other modules) and FEM losses, the available SNR for QPSK signal at LTE 20 MHz BW is ~2 dB [(−94) − 5.5 − (−101.4)].

As presented in the next chapters, many components, such as I-Q imbalances, nonlinearity, quantization, signal interferers/blockers create multiple components of unwanted signal that add up with the receiver noise floor. As an example, for an unwanted interference, that is, 6 dB below the receiver noise floor, the combined effect can degrade the SNR by 0.97 dB (table 3.3). Similarly many such unwanted components added up can further degrade the signal quality—even if they are lower than noise power level by several dBs.

4 LTE FREQUENCY AND CLOCK SOURCES

The reference clock for the long-term evolution (LTE) system is one of the key items to achieve signal quality. This and the next chapter deal with clock sources for LTE, frequency errors, and their relation with signal quality. A functional representation of eNodeB (eNB) and user equipment (UE) is shown in figure 4.1. As marked in figure 4.1, the eNB maintains the master frequency (or timing) reference to the absolute accuracy of ±0.05 parts per million (ppm or PPM), or 50 parts per billion (ppb) or better [URL (gps-gov), 3GPP 36.104]. Note that the ppm and ppb are in ± values. The UE is required to acquire the eNB frequency reference through received downlink (DL) signal.

To illustrate with an example, the UE hardware uses either a crystal oscillator (XO) with ±10 to ±20 ppm (as an example ppm) or temperature compensated XO (TCXO) with ±0.25 to ±1 ppm. As per 3GPP 36.521-1, 3GPP 36.101 requirements, the uplink (UL) carrier has to be within ±(0.1 ppm + 15 Hz) accuracy with reference to DL carrier. (Note: To keep the text in compact form, ±0.1 ppm or 0.1 ppm is used in rest of the chapters in place of ±(0.1 ppm +15 Hz). In the presence of relative velocity between eNB and UE, the Doppler can deviate by more than ±0.1 ppm, and examples on this topic are given in chapter 5.

Initially the UE receiver local oscillators (LO) operate with the arbitrary ppm error of XO or TCXO. The frequency error estimation and frequency tracking inside the UE continuously (at regular update intervals) derive relative error between internal reference and the received DL carrier. The relative error is applied to frequency synthesizers, and the synthesizers adjust the frequency matching close to DL carrier. The received carrier can have Doppler shift when UE is in relative motion with reference to the eNB, and UE tracks the received carrier to match the Doppler-shifted carrier. Thus the use of a high-precision clock source, such as oven-controlled XO (OCXO), alone will not be good enough inside the high-speed moving UE without frequency tracking. The preferred approach then is to use frequency-tracking techniques operating on the received DL signal.

Once frequency tracking is incorporated in UE, TCXOs or lower ppm crystal oscillators may be used.

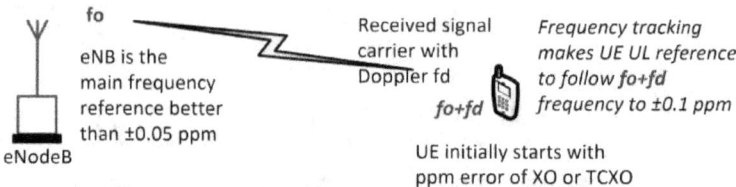

Figure 4.1. Frequency reference representation at UE.

4.1   Clock Sources for LTE Operation

The popular clock sources used in user equipment (UE) are given here:
- Crystal oscillator (XO)—passive crystal with Pierce oscillator.
- Temperature-controlled crystal oscillator (TCXO).
- Precision clock reference generated internal to the system, such as from global positioning system (GPS).
- Oven-controlled crystal oscillator (OCXO)—mainly applicable to femtocells and base stations.

4.1.1   Crystal Oscillator

The implementations of crystal oscillator possibilities are many [URL (microchip)]. In this section, examples are considered with Pierce oscillator. As shown in figure 4.2, the inverter gate introduces 180 degrees phase shift, and the passive network along with crystal introduces another 180 degrees phase shift, making the circuit to oscillate. The shunt resistor R_F turns square waves into smooth transitions. The shunt capacitors C1 and C2 influence the adjustment of frequency. Within the valid frequency range, the lower value of shunt capacitance increases the center frequency and the higher value of capacitance decreases the center frequency. The configuration in figure 4.2(a) is popularly used with digital chips, and the second configuration shown in figure 4.2(b) is more common with analog, mixed-signal and radio-frequency (RF) chips. The main difference between figures 4.2(a) and 4.2(b) being the resistors R_S and R_F, and the fixed or variable capacitors go inside the RF/analog chip. By varying these capacitors through programming, crystal center frequency can be tuned at a coarse (big steps) level. Taking the example of crystal oscillating starting at ±10 ppm deviation, during the LTE connection coarse frequency estimation phase, the capacitance value can be altered to reduce the initial ppm offset. The capacitance adjustments help such first-order coarse corrections. A crystal oscillator adjusted close to the center frequency behaves better in stability. The fine adjustments can be propagated to synthesizers, which are explained in later sections. Figure 4.2(b) is for varying capacitance to correct the center frequency. Given a ±10 ppm crystal, upon activation of LTE, a coarse adjustment can be made with programming variable capacitors $C1_V$ and $C2_V$ to achieve approximately ±2 to ±5 ppm. The capacitors $C1_f$ and $C2_f$ are external fixed values for initial start-up of the crystal oscillator.

The crystal oscillator ppm behavior can also vary due to reflow at soldering. The oscillator frequency varies depending on many factors, such as temperature, aging (of crystal and passives), humidity, vibration, supply voltage variation, and loading. Out of all these parameters, temperature is the main parameter that quickly alters the crystal characteristics and results in ppm variation. If temperature data are known, the coarse-level temperature adjustments may be taken care of with capacitance adjustments.

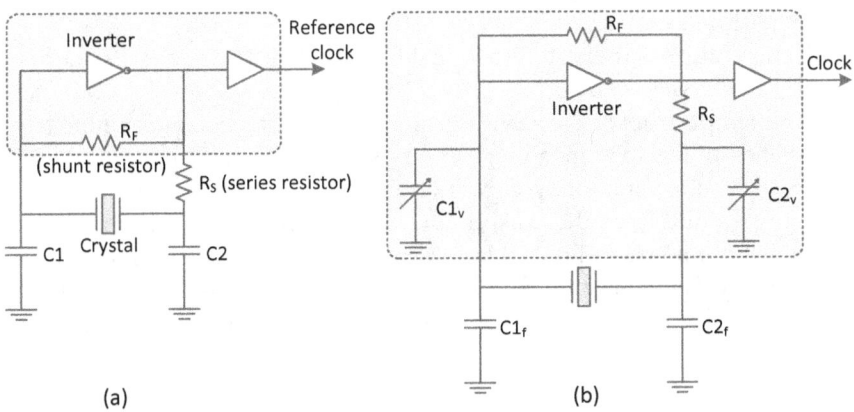

Figure 4.2. (a) Crystal oscillator with digital chip. (b) Crystal oscillator with RF/analog chip.

4.1.2   TCXOs

In the temperature-controlled crystal oscillator, temperature-dependent compensation is performed that keeps crystal oscillator ppm to a narrow range of ±0.25 to ±1 ppm versus the crystal oscillator frequency tolerance of ±10 to ±20 ppm. While fabricating, TCXOs are made with tuning and frequency adjustments. TCXOs are commonly used in UE devices.

A functional representation of TCXO is shown in figure 4.3 [URL (Vectron-TCXO)]. Once a particular cut crystal is selected, several parameters can influence the crystal oscillator frequency. As stated earlier, the main parameters that vary the crystal frequency are temperature, voltage, parasitic capacitances, shock, and vibration. Inside the TCXO, the input supply voltage is regulated and precision supply is used with the oscillator, and this eliminates supply variation dependency.

Functionally variable capacitors are provisioned in place of figure 4.2(a) C1, and C2 of the crystal oscillator. These capacitors help in setting the oscillator ppm to approach close to zero at the nominal temperature of +22 °C to +25 °C. On the basis of the sensed temperature, a temperature sensor (a set of thermistors) generates the voltage. The voltage is properly conditioned through filters and gain scaling. A few common overlapping options with the temperature information usage are described here [URL (Crystek-TCXO), URL (Vectron-TCXO)].

Voltage goes to variable capacitors (varicaps) that changes the capacitance and in turn modifies the ppm drift. The voltages and varicaps will be such that to work in opposite direction of ppm drifts. In advanced implementations, a digital memory table is built that stores the temperature-dependent compensation. The common approaches to vary the capacitance are further listed below:

- Temperature is translated to digital bits and bits switch the capacitors [URL (Crystek-TCXO)].
- The temperature sensor analog voltage after suitable conditioning is applied to the varicap, and varicap changes the capacitance and thus regulates the variations to lower limits of ±0.25 to ±1 ppm.
- Temperature from the sensors such as thermistors is converted to digital numbers, and digital numbers drive the memory lookup. The memory values are programmed with good curve fitted parameters for maintaining minimal ppm (or maximum correction).

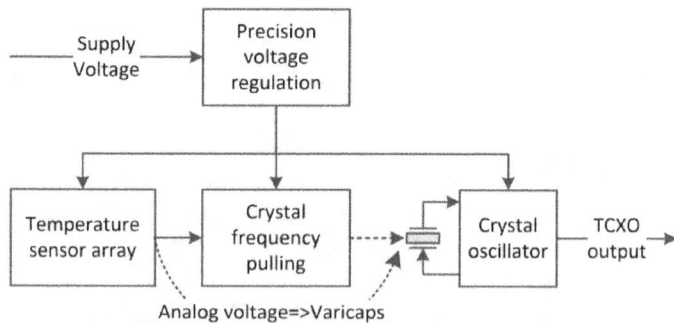

Analog voltage=>Varicaps
Digital bits=>Caps switching and analog voltage to varicaps
Digital bits=>Memory=>Caps switching and analog voltage to varicaps

Figure 4.3. TCXO functional blocks.

As TCXOs are in a shielded package and output is buffered, the parasitic capacitances and electrical load may not influence frequency change, and TCXO package also partly helps absorb shock and vibrations from the usage environment.

4.1.3   OCXO

In OCXO, the temperature is held constant much above the room/ambient temperature. A heater, sensor, and control loop keeps the oven temperature constant. A well-tuned crystal, voltage control, and constant temperature maintain OCXO oscillator frequency to better than ±0.010 to ±0.050 ppm [URL (Rakon-OCXO)]. In some OCXOs double oven is used to further enhance the precision/stability. OCXO is relevant to stationary air interfaces, such as LTE base stations, femtocells, and instruments, and not applicable to LTE UE. When UE is in motion at more than 108 km/h (the math on velocity, Doppler, and ppm relation are given in chapter 5), the Doppler can be more than the acceptable UL 0.1 ppm carrier frequency deviation that demands use of frequency errors estimation and closed-loop DL frequency tracking. The real benefit of ultrastable OCXO is not fully helpful with high-mobility UE.

4.1.4   GPS Derived Clock

GPS systems are for position information. GPS systems can also be used for deriving the precision clock. If GPS is reliably deriving the precision clock, such clock can also be used with LTE. GPS derived precision clock references are used in femtocells and home eNBs [URL (Symmetricom)].

4.2   Frequency Synthesizers

The main purpose of frequency synthesizer is to take XO or TCXO reference signal/clock and generate the required transmit (Tx) and receiver (Rx) LO reference and high-frequency clocks to mixed-signal modules. The LTE LO frequency is different depending on the band of operation and needs to be adapted as per channel raster of 100 kHz for all bands [3GPP 36.521-1]. Channel raster 100 kHz means the carrier center frequency is integer multiple of 100 kHz.

The synthesizer working with default XO or TCXO reference will not be sufficient for LTE accuracies. In this section, assuming LTE receiver's frequency-tracking block (signal processing algorithms present in baseband) is estimating the error with reference to the Rx signal (carrier) center frequency, the correction mechanism blocks are shown in figure 4.4. In some UE designs, the crystal oscillator or TCXO feeds clock to one common low frequency synthesizer. The synthesizer can have correction loop from LTE receiver. The corrected master clock works as base reference to use with multiple synthesizers. In figure 4.4, separate synthesizers are assumed to be used with Rx synthesizer, Tx synthesizer, and analog-to-digital converter (ADC) and digital-to-analog converter (DAC) sampling clocks.

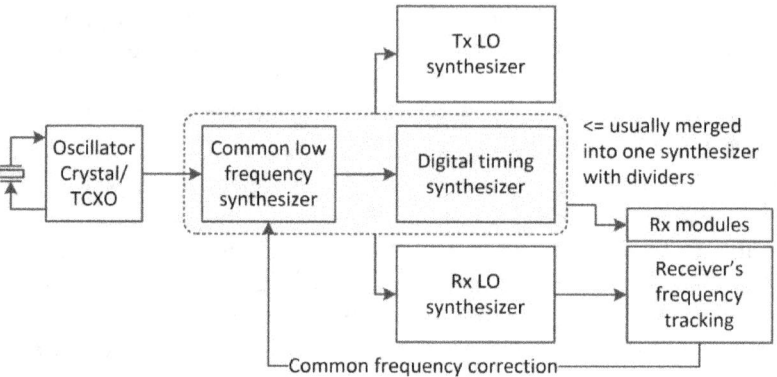

Figure 4.4. Crystal oscillator or TCXO used with common synthesizer.

4.2.1   Distinct Tx, Rx, and Clock Synthesizer Correction

As shown in figure 4.5, the clock from crystal oscillator or TCXO goes to multiple synthesizers, and each synthesizer gets correction programmed independently [URL (LMS7002M)].

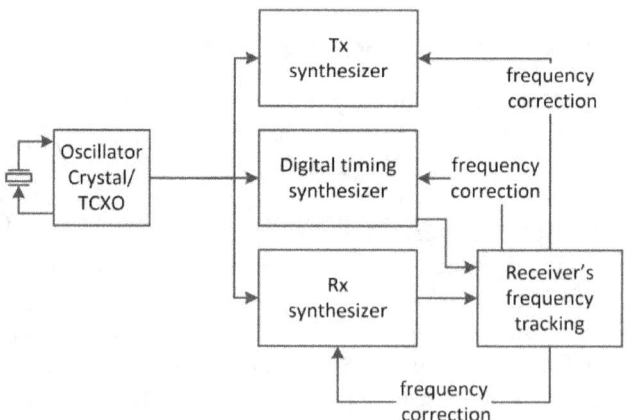

Figure 4.5. Crystal oscillator or TCXO driving separate synthesizers.

The independent synthesizer approach gives more freedom to tune each synthesizer differently. Using this approach the digital timing synthesizer can be tuned differently to account for sampling clock offsets (SCO) as described in chapter 5. The LTE signal processing algorithms generate SCO that has to be decoupled with LTE Rx and Tx local oscillator frequencies.

4.3   Int-N and Frac-N Synthesizers

An integer (Int)-N synthesizer takes f_{ref} and provides N times the f_{ref}, where integer value N is programmable to get variation in frequency. There could be predivider to arrive at f_{ref} and postdivider on voltage-controlled oscillator (VCO) frequency f_{vco} to derive multiple frequency combinations (figure 4.6).

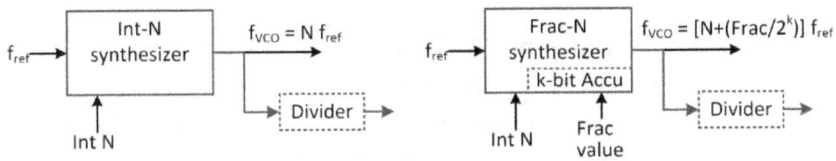

Figure   4.6.   Integer-N   and   Fractional-N   synthesizer   functional representation.

4.3.1   Frac-N Synthesizers

Frac-N synthesizer takes fractional part and integer part and generates VCO frequency as given by

$$f_{VCO} = \left(N + \frac{Frac}{2^k}\right) f_{ref} \tag{4.1}$$

$$f_{VCO} = \left(N + \frac{0}{2^k}\right) f_{ref} = (N) f_{ref}. \tag{4.2}$$

The Frac value is from 0 to 2^k-1, and divider resolution is $f_{ref}/(2^k)$. For k of 24 bits, f_{ref} of 26 MHz, the resolution is given as 26 MHz/(2^{24}) = 1.55 Hz. The resolution step error is half of resolution ±1.55/2 = ±0.775 Hz. The input may be going through divider by R and generating f_{ref}. The f_{vco} may be used with postdivider for the required scaled down frequency. In figure 4.7, k-bit accumulator is shown with the Frac value. Figure 4.7 is a functional representation to convey the principle, and the actual implementations use sigma-delta modulators with dividers [URL (MIT-Synth), URL (ti-Synth), URL (linear-Synth), URL (Skyworks)] and many advanced techniques to improve phase noise and spectral spurious components. The voltage-controlled oscillator (VCO) frequency goes through fractional divider and compared with f_{ref}. The phase frequency detector (PFD) produces error current/voltage on the basis of the frequency and phase difference that controls the VCO frequency in a closed loop.

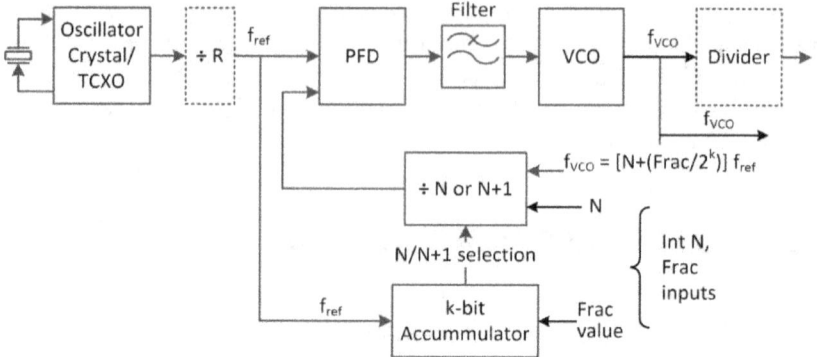

Figure 4.7. Fractional-N synthesizer basic principle diagram.

4.3.2   Frac-N Synthesizers Resolution for LTE Bands

Considering LTE lower frequency of 751 MHz for Band-13 (B13) DL, the ppm error due to frequency resolution of 1.55 Hz is $(1.55/2)/751 = 0.001$ ppm or 1 ppb. The LTE maximum allowed error is 0.1 ppm, and 0.001 ppm is 1/100th of 0.1 ppm. If $k = 20$ bits, resolution is 26 MHz/(2^{20}) = 24.8 Hz, and error at 751 MHz is 0.016 ppm or 16 ppb, which is a big step of maximum allowed 0.1 ppm in the UL. Table 4.1 lists the percentage (%) error due to synthesizer resolution. For LTE UE applications, 22 or 24 bits are a more suitable selection to create correction resolution to be a small fraction of 0.1 ppm.

Table 4.1. Frac-N examples for resolution and ppm error for Band-13 and Band-7 (B7)

f_{ref} MHz	k-bits	Resolution $(f_{ref}/2^k)$ in Hz	UL max allowed error ±ppm	±step ppm error at B13 782 MHz	±step error% with reference to max ±0.1 ppm	±step ppm error at B7 2,535 MHz	±step error% with reference to ±0.1 ppm
26	20	24.80	0.1	0.016	16%	0.0049	4.9%
26	22	6.20	0.1	0.004	4%	0.0012	1.2%
26	24	1.55	0.1	0.001	1%	0.0003	0.3%
52	20	49.59	0.1	0.032	32%	0.0098	9.8%
52	22	12.40	0.1	0.008	8%	0.0024	2.4%
52	24	3.10	0.1	0.002	2%	0.0006	0.6%

Notes: As UE is a slave device of eNB, the resolution is required to track eNB reference. The eNB starts with precise reference of better than 0.05 ppm with raster of 100 kHz, and often a lower resolution frac value is used with eNB, but higher-resolution frac values are suggested for UE. Higher resolution refers to a smaller value of ($f_{ref}/2^k$).

4.4   Phase Noise and Jitter

Crystal oscillators or TCXOs produce phase noise, and phase noise can also be represented as jitter. Phase noise is the frequency response of the unwanted leakage around the center frequency, and jitter is the time shift around zero crossings of the sine wave or square wave [URL (nist1), URL (ADI1)]. When crystal oscillator reference is used in generating RF LO frequency, the phase noise is passed on to integer (int) or fractional (frac) synthesizers. The integer synthesizer takes reference frequency and generates integer multiples of the input reference frequency. With the LTE system requirements, integer multiples of the reference are not sufficient (to achieve 0.1 ppm and 100 kHz raster resolution for multiple bands coverage), and fractional synthesizers are used.

As explained in the previous sections, the crystal oscillator reference can drift in frequency, and fractional synthesizer is helpful in correcting the frequency to a few hertz (Hz) resolution. Crystal oscillator output may have random and deterministic amplitude and phase noises [URL (Crystek)] and various frequency drifts explained in the previous sections. For the purpose of mobile system signal quality, the main concerns are with short-term (hundreds of microsecond scale) phase noise that limits SNR, and frequency drift that rotates constellation points within the slot/subframe.

A thorough analysis and interpretation on phase noise in signal sources is available in reference [Rabins (84)]. Random phase noise due to short-term frequency instability is represented as a spectral domain plot [URL (Rakon1), URL (Rakon2)]. In the LTE system, the stable references are used at ADC/DAC to interface baseband digital samples, and Rx/Tx LOs to down or up-convert to RF signals. For ADC/DAC mixed-signal analysis, jitter is used in relation to sampling clock, and jitter-based math reveals the signal-quality issues. For RF-LO, spectral domain analysis is used treating LO as a sinusoid signal. The relations among phase noise and jitter are well established [URL (ADI1), URL (Rakon1)].

4.4.1   Phase Noise Frequency Regions

In the literature, phase noise is shown in several valid ways. In reference [URL (ADI1)], the phase noise is shown as close-in phase noise that limits the frequency resolution and far-away broadband noise that limits signal-to-noise ratio (SNR). For the LTE system, the frequency (tone) resolution is 15 kHz; hence, close-in frequency may not cause significant issues, and the total integrated phase noise influences the SNR of the signal analysis.

The phase noise may be modeled as multiple frequency segments as marked in table 4.2 [URL (Nist1), URL (Rakon2), URL (ADI1)]. In sensitive applications such as space communication, the precise locations of these segments are important. For LTE signals, the approximations are good enough, and the spectrum analyzer readings are approximated to decade segments. The typical frequency corners for analysis starts with 10 or 100 Hz, which typically skips part of the random-walk ($1/f^4$) region close to the center frequency. Part of the issue with random walk is the measurement capability of the instruments. Refer to [Robins W.P. (1984)] for a good overview on phase noise and various formulations. The measurements include a few selected segments usually in the decades scale, such as 100 Hz, 1,000 Hz, 10 kHz, 100 kHz, 1 MHz, and so on. As illustrated in [URL (Rakon1)], the phase noise is measured in decade frequency steps. The instruments [URL (Keysight-PN)] can provide the integrated phase noise, or mathematically these segments can be used to calculate the total phase noise.

Table 4.2. Phase noise regions and their description

Phase noise category name	Description and parameters influencing phase noise	Slope	Slope dBs/decade
Random walk	Contributed from multiple sources—quartz structure, short term mechanical shock, vibration, temperature changes etc.	$1/f^4$	−40
Flicker frequency	Intermodulation of white frequency carrier noise and flicker phase from buffers	$1/f^3$	−30
White frequency	Carrier noise, crystal RLC (resistor, inductor, and capacitor) noise	$1/f^2$	−20
Flicker phase	Pink noise from buffers and transistors	$1/f$	−10
White phase	Thermal noise (flat noise response)	$1/f^0$	0

4.5   Phase Noise and Nonlinearity

Clock and frequency error characteristics are given in chapter 5, and in this section the connections between Tx and Rx signal levels, Tx and Rx local oscillators (LOs), blockers, and phase noise are summarized.

Tx phase noise and nonlinearity:
- Refer to [URL (Aeroflex-7100), URL (R&S-1), URL (Anritsu-8821C), URL (NI-testing), 3GPP 36.521-1] for list of various impairments and measurements.

- Tx up to the mixer up-conversion input stage is reasonably linear to the extent that it will not influence Tx signal quality through phase noise.
- At the Tx up-conversion mixer stage, the two key components of distortions are the LO phase noise and nonlinearity or intermodulation (IM) distortion.
  - As the mixing operation is a convolution in the frequency domain, the double side band LO phase noise gets convolved with every orthogonal frequency-division multiplexing (OFDM) tone and limits the signal to noise ratio (SNR).
  - The intermodulation in the mixer stage—if not taken care—can appear as in-band distortions and also as adjacent channel leakage.
- The Tx PA is the key contributor of intermodulation, and the IM level depends on various contributors, such as input PAPR, output power level, gain/attenuation selection, input/output matching, V_{BAT} voltage, and the usage of fixed, average-power, and envelope-tracking voltage supplies with various signal levels and gain/power control.

Rx phase noise and nonlinearity:

- In FDD, Tx and Rx work simultaneously with tens of MHz spectral separation [URL (free), 3GPP 36.101]. The Tx signal and its nonlinearity components leak into Rx through duplexer (and FEM), and Tx spectral side lobes can compete (work as blockers) with the Rx desired signal.
- The blockers signal entering on Rx side and Tx leakage entering through duplexer can create intermodulation at LNA.
- The Rx LO double side band phase noise establishes the limits on received signal SNR.
- The Rx side blockers signal and Tx leakage can create both phase noise and intermodulation contributions in mixer down-conversion stage.
- In the LTE TDD, the above topics are still applicable except for the Tx leakage. In TDD Tx leakage will not overlap in time with Rx signal.

4.6   Jitter Math from Phase Noise

For a typical data example given in [URL (Rakon2)] for 13 MHz, the basic operation are shown in table 4.3. The operations here are first-level approximations to convey basic math in arriving at phase noise and jitter. Interpretation of table 4.3 math is listed below:

- Frequency is shown in row Hz, and the corresponding dBc value in row dBc.
- Segment A1 power is the average phase noise power = (power at f_1 + power at f_2)/2.

- Frequency span of A1 segment is in dB scale = $10 \log_{10}(f_2 - f_1)$.
- Integrated A1 segment noise power in dB units = average noise power from A1 + $10 \log_{10}(f_2 - f_1)$.
- Jitter in radians from A1 = $\mathrm{sqrt}(2 \times 10^{(\text{integrated A1 segment power}/10)})$.
- Jitter_A1 in seconds = (Jitter radians from A1)/$(2 \pi f_0)$, where f_0 is the oscillator center frequency, and in the table (ps) is picosecond units.
- Total rms jitter from all the segments $t_j = \mathrm{sqrt}(\text{jitter\_A1}^2 + \text{jitter\_A2}^2 + \cdots + \text{jitter\_A}n^2) = 0.6316$ ps.
- The SNR limit in dBs on signal at frequency f_0 (here 13 MHz oscillator frequency) is given as $-20 \log_{10}(2 \pi f_0 t_j)$;
  $-20 \log_{10}(2\pi \times 13 \times 10^6 \times 0.6316 \times 10^{-12}) = 85.75$ dB.

Table 4.3. Jitter calculation

Spot name	f_1	f_2	f_3	f_4	f_5	f_6
Frequency Hz	10	100	1,000	10,000	100,000	1,000,000
Power dBc	−100	−128	−142	−148	−150	−151
Segment name		A1	A2	A3	A4	A5
Average phase noise dBc		−114	−135	−145	−149	−150.5
$10 \log(f_{n+1}-f_n)$		19.542	29.542	39.542	49.542	59.542
Integrated segment noise power in dB		−94.457	−105.457	−105.457	−99.457	−90.957
Jitter in segment (radians)		2.676×10^{-5}	7.544×10^{-6}	7.544×10^{-6}	1.505×10^{-5}	4.005×10^{-5}
Jitter in segment (seconds)		3.27×10^{-13}	9.23×10^{-14}	9.23×10^{-14}	1.84×10^{-13}	4.90×10^{-13}
Jitter in segment (ps)		0.3277	0.0923	0.0923	0.1842	0.4903
Square of jitter (ps^2)		0.1074	0.0085	0.0085	0.0339	0.2404
Sum of (ps^2)		0.3989				
Total integrated RMS jitter in ps		0.6316				
SNR in dB		85.75				

From table 4.3, the dominant jitter is appearing from segment A5 due to contributions from wide frequency span from 100 kHz to 1,000 kHz as $10 \log_{10}(f_6 - f_5)$. Even though frequency span is low for segment A1, the phase noise power at f_1 and f_2 are higher; hence, jitter contribution is also higher. The jitter SNR is appearing as close to 86 dB. When such crystal is used with the LTE system for center frequency of, say, 2,600 MHz, the SNR will degrade by 20 $\log_{10}(2,600/13) = 46$ dB, and this varies slightly depending on how the 2,600 MHz is generated using reference 13 MHz. As explained in chapter 2, the LTE system typically uses 26 MHz that helps in creating some more improvements to the final phase noise numbers.

Notes: The segment-based calculation is an approximation, and for precise math, instruments may be used, or more segments can be used. The SNR in the examples $86 - 46$ dB $= 40$ dB is a very good value even to use 13 MHz for the LTE system.

4.7   Jitter Influence on Sampled Signals SNR

As referenced in [URL (ADI2)], ADC aperture jitter and sampling jitter limit SNR. The jitter has to be much lower to achieve higher SNR. To clarify further on the same aspect, the noise floor due to jitter has to be much below the quantization noise floor.

Aperture delay is the relative delay variation between the sampling clock edge, and the actual analog signal sampling instant. The variation in aperture delay between in-phase and quadrature-phase (I-Q) samples create time delay or linear phase distortion. The sample-to-sample aperture delay variation is the aperture jitter. By design, the aperture jitter is kept very low, and its contributions are made to much lower than the quantization noise.

Let t_a is the aperture jitter, the SNR limit in dB on signal input at frequency f_{sig} is given as $-20 \log_{10}(2\pi f_{sig} t_a)$. For an aperture jitter t_a and clock jitter or clock-skew jitter t_j, the combined rms jitter is $t_{ja} = \text{sqrt}(t_a^2 + t_j^2)$, and SNR limit is given as $-20 \log_{10}(2\pi f_{sig} t_{ja})$.

4.8   Power Dissipation, Voltage, and Temperature

The power dissipation, voltage, and temperature are hardware device specific and influence mainly RF and analog sections. As explained in the previous chapters, the crystal oscillators drift (changes frequency) with temperature, voltage, and capacitance. The gain and DC offset of amplifiers vary with supply voltage and temperature. The periodic jitter/noise originated due to power supply ripple and ground (shared common path internal to the chips) noise—usually ignored—can influence phase noise and spectral spurs. Depending on the activity of Tx and Rx, the power dissipation varies for multiple chips/devices, and such variation can modulate signals during subframe intervals. During inter–radio access technology (RAT) and intra-RAT, the power dissipations can vary, which will have influence. During inter-RAT when power dissipation is varying, the ripple or noise on power supplies can vary and can influence the phase noise and spectral spurs. The spurs and phase noises optimized for one band or one RAT may not be fully valid in another band or another RAT. Refer to [URL (Radisys-RAT)] for intra- and inter-RAT call flows. (Note: As a simple example, inter-RAT is leaving the 4G-LTE network and entering the 3G or 2G network. The intra-RAT is switching from one eNB to another or switching from, say, LTE Band 13 to a different Band 4.)

4.9   Summary on LTE System Clock Generation

The focus of the chapter is on UE reference clock that creates signal-quality degradation mainly through Tx LO, Rx LO, and mixed-signal ADC/DAC stages. As a common practice, the LTE physical-layer time-domain processing is synchronized to mixed-signal clock. All other processing clocks inside UE need not be precise and need not be in synchronization with the main reference clock. The TCXO is popularly used in systems that can absorb more expensive bill of material (BOM), and for lower BOM products ± 10 ppm (or better) crystal oscillators are used with multiple techniques to improve ppm. The lower the ppm, the stable the system, and it helps signal quality. The frequency-tracking algorithms work better if the ppm error is lower for correction. In the UE, using a precise oscillator, such as OCXO may not help as UE can receive Doppler-shifted Rx carrier that may be shifted by more than ± 0.1 ppm when UE is at more than 108 km/h. The Tx and Rx LO frequencies have to cover in steps of 100 kHz raster, and also frequency synthesizers have to track finer frequency corrections that enforce the need for frac-N synthesizers.

The main clock reference frequencies for LTE-UE are 26 or 52 MHz. In theory any oscillator with frac-N synthesizers may be used. The phase noise of the reference low frequency oscillator is one of the prime noise creator through synthesizers and LOs. As an example, assuming Tx/Rx LO frequency is N times the reference, then phase noise increases by approximately $20 \log_{10}(N)$ or more. The higher the frequency reference, the lower will be the N to generate the same Tx/Rx LO frequency, and better will be the phase noise. As an example, the clock oscillator 13 or 26 MHz will have almost similar phase noise characteristics, but using 26 MHz gives benefit to the final RF synthesizer phase noise. In the preference order, 26 MHz is preferred to 13 MHz reference, and 52 MHz is preferred to 26 MHz. The main preference for 26 or 52 MHz comes from the mobile industry processor interface (MIPI) explained in chapter 2. These interfaces are made to work with 26 MHz references; hence, the use of 26 or 52 MHz oscillator makes sense.

5 FREQUENCY ERRORS AND SIGNAL QUALITY

Orthogonal frequency-division multiplexing (OFDM) signals are framed with time and frequency parameters [Moray Rumney (2013), Chris Johnson (2012), Christina Gebner]. An OFDM signal has multiple tones with each tone of 15 kHz frequency resolution that relates to the reciprocal of time window of 66⅔ μs (microseconds) referred to in this book as 66.66 μs duration (chapter 1). A rectangular window of 66.66 μs duration shown in figure 5.1(a) is a sinc—here sinc is $\sin(x)/x$—function in the frequency domain as given in figure 5.1(b). A tone at frequency 15 kHz with rectangular time window is shown in figure 5.1(c). As shown in figure 5.1(d), in the frequency domain, this is a sinc function centered at 15 kHz. The OFDM symbol has a varying number of tones depending on the LTE bandwidth (BW). LTE 1.4 MHz BW uses 72 tones, and 20 MHz uses 1,200 tones. Considering LTE BW of 20 MHz, under ideal conditions the OFDM tones are placed from index −600 to 600, and figure 5.1(e) is represented for just 3 tones. These tones are orthogonal to each other—and the meaning for OFDM is explained in several valid ways [Stefania Sesia et al. (2011), Award Solutions]:

- The 66⅔ μs window holds an exact integer number of cycles for tones from index −600 to 600, and each tone is spaced 15 kHz apart.
- In the frequency-domain processing using fast Fourier transform (FFT) or inverse FFT (IFFT), every tone appears in the sinc function zero crossing positions of all other tones as shown in figure 5.1(e).

Distortions in orthogonality appear when the following happens:

- Sampling is not creating exact 66⅔ μs symbol duration, which means for LTE 20 MHz BW the sampling is not exact 30.72 MHz.
- The analog-to-digital converter (ADC) analog input tones are not spaced at 15 kHz.
- Input analog signal is frequency shifted.

With frequency errors or distortions in orthogonality, the tones undergo amplitude loss and intercarrier interference (ICI) that translates into signal-to-noise ratio (SNR) loss as marked in the next sections.

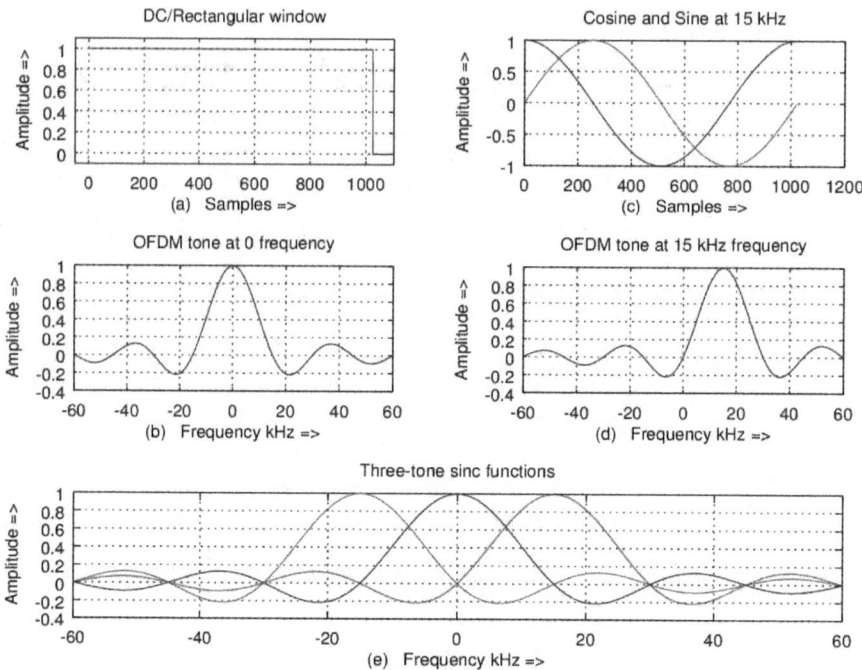

Figure 5.1. (a) Rectangular window. (b) OFDM tone at 0 frequency—same as frequency response of rectangular window as the sinc function. (Note: In parts [a] and [b] the x axis is marked as samples, and in this example a block of 1,024 samples creates 1-LTE symbol of 66.66 μs.) (c) Tone with rectangular window. (d) Sinc function centered at 15 kHz. (e) Multiple sinc functions showing orthogonality characteristics.

> Notes: The tone (subcarrier) spacing is 15 kHz, and the half subcarrier spacing is 7.5 kHz. In the uplink, the tones are half subcarrier shifted and starts at −7.5 kHz and +7.5 kHz by maintaining the same 15 kHz spacing. In the downlink the center subcarrier is discarded.

5.1   Doppler Shift in LTE System

The following summary points convey the overview and basic math with examples on Doppler frequency shift (figure 5.2):

- Doppler shift in frequency is created because of relative velocity or motion between the base station or eNodeB (eNB) and user equipment (UE), uplink (UL) and downlink (DL) frequency, and the angle of the UE velocity vector with reference to the base-station antenna.

- As the eNB is stationary, the Doppler frequency is created by the velocity of the UE.

- The Doppler frequency in hertz is given as $(V_r/\lambda) \cdot \cos(\theta)$, where V_r is the velocity in meters per second (m/s), θ is the angle of velocity vector with reference to the base station, and λ is the wavelength given as (c/f_o) in the same meter units, c is the velocity of light $\sim 3 \times 10^8$ m/s, and f_o is the center frequency of radio-frequency (RF) signal in hertz.

- When UE is directly moving away or approaching toward the base station, $\cos(\theta) \approx 1$. The UE traversing across (tangential) makes $\cos(\theta)$ approximately zero. In table 5.1, the Doppler is shown positive (increase in frequency), which means the UE is approaching toward base station. If UE is going away from the base station, the Doppler is negative (decrease in frequency).

- Taking an example of LTE Band-4 (B4) DL at 2,132 MHz (λ of 0.1407 m), and 500 km/h (V_r of 138.9 m/s) high-speed (HS) train, the Doppler frequency (V_r/λ) is 987 Hz, and for UE going away from the base station, the Doppler frequency is −987 Hz. In terms of parts per million (ppm or PPM) contribution, the approaching UE creates $987/2,132 = 0.46$ ppm.

- LTE Band 13 (B13) gives lower Doppler frequency, and the relative ppm is still the same as in B4. As V_r/λ is the same as $V_r/(c/f_o)$ and normalized with f_{o_MHz} for ppm, it gives $V_r/(c/f_o)/(f_{o\_MHz})$, which gives quantity proportional to V_r/c making it ppm error/drift independent of f_o.

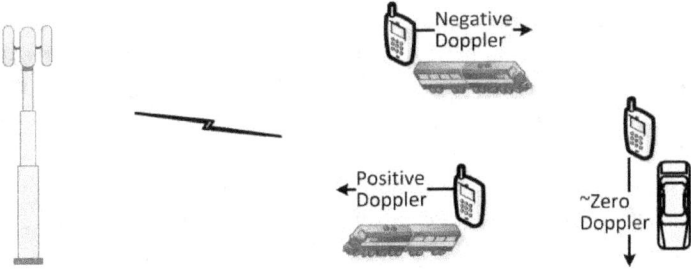

Figure 5.2. Doppler shift functional representation with the moving UE (moving away, coming toward, and moving tangential).

Table 5.1. Maximum Doppler frequency at multiple velocities and band frequencies with cos(θ) = 1

Velocity		B13 DL (751 MHz) Doppler parameters				B4 DL (2,132 MHz) Doppler parameters			
km/h	V_r in m/s	B13 f_0 in MHz	λ in m	Doppler V_r/λ in Hz	Doppler shown as ppm	B4 f_0 in MHz	λ in m	Doppler V_r/λ in Hz	Doppler shown as ppm
10	2.78	751	0.40	6.95	0.01	2,132	0.14	19.74	0.01
50	13.89	751	0.40	34.77	0.05	2,132	0.14	98.70	0.05
100	27.78	751	0.40	69.54	0.09	2,132	0.14	197.41	0.09
108	30.00	751	0.40	75.10	0.10	2,132	0.14	213.20	0.10
150	41.67	751	0.40	104.31	0.14	2,132	0.14	296.11	0.14
200	55.56	751	0.40	139.07	0.19	2,132	0.14	394.81	0.19
250	69.44	751	0.40	173.84	0.23	2,132	0.14	493.52	0.23
300	83.33	751	0.40	208.61	0.28	2,132	0.14	592.22	0.28
350	97.22	751	0.40	243.38	0.32	2,132	0.14	690.93	0.32
400	111.11	751	0.40	278.15	0.37	2,132	0.14	789.63	0.37
450	125.00	751	0.40	312.92	0.42	2,132	0.14	888.33	0.42
500	138.89	751	0.40	347.69	0.46	2,132	0.14	987.04	0.46
		B13 UL frequency = 782 MHz				B4 UL frequency = 1,732 MHz			

5.2   Frequency Offset Estimation

Frequency offset error or correction value is referred to by its common name—carrier frequency offset (CFO). The CFO can be further split into integer frequency offset (IFO), fractional frequency offset (FFO), and residual frequency offset (RFO) [Stefania Sesia et al. (2011), URL (dline-CFO), URL (nctu)]. In general several closely matching names are used to convey the same meaning. The IFO is called coarse frequency correction, and FFO and RFO together are called fine frequency correction. In LTE system, IFO is the integer multiples of 15 kHz correction, and FFO is residual part less than 15 kHz. The naming usage is not strict—beyond the 3GPP standards—and need to be interpreted depending on the context of usage.

LTE Frequency error can be estimated using multiple ways, and most common approaches are based on primary synchronization signal (PSS) for IFO and FFO, and pilot tones for RFO. The possibilities of frequency estimation are many depending on the overall system parameters, and digital signal processing techniques used. In the subsequent parts of the section, factory calibration, cyclic prefix (CP), PSS half symbol–based estimation, and pilot tones–based approaches are summarized at a high-level.

Factory calibration. The preferred way to estimate coarse-level frequency error is during the factory calibration. As a part of unit level testing of the UE, the UE oscillator is of free running with crystal oscillator or TCXO. The factory calibration can estimate the coarse frequency error at room temperature. The error can be stored in the UE flash memory and applied at next time power on either to correct reference oscillators or fractional synthesizers. With this approach UE can wake up at less ppm error of the order of ±1 to ±5 ppm. The factory calibration when performed tightly can avoid the necessity of IFO.

PSS-based IFO. PSS is generated using frequency-domain Zadoff-Chu (ZC) sequence [URL (Mathworks-PSS)]. It has 62 tones with 31 tones on either side of DC subcarrier. These tones are correlated in the receiver with multiple ZC sequence frequency-domain templates (tone modulations). When the sequence correlation matches, the frequency lag (shift in correlation) gives the IFO—in terms of the number of tones shifted, and each tone shift is of 15 kHz.

PSS-based FFO. Once IFO is corrected in the receiver, the cyclic prefix–based approach can be used to obtain fractional frequency offset. The common approach is to use CP-based correlation of PSS as given in [URL (nctu)]. The PSS-CP is part of the PSS symbol and precedes the symbol. The CP is correlated with the symbol that gives relative phase shift at the peak. The CP is 1 symbol length away. Using the phase shift in 1 symbol duration, frequency can be estimated. As CP is of shorter duration ~5 μs, the half symbol duration (33.33 μs) based correlation gives better estimation. In PSS, the first half of the time-domain symbol has conjugate relation with the second half. This gives longer duration correlation window of 33.33 μs and hence improves accuracy for FFO frequency estimation. The half symbol–based correlation method is reported in [URL (nctu)].

RFO with pilot tones. The reference signals (pilot tones) are present as parts of physical uplink shared channel (PUSCH) symbols. By observing phase changes on the specified reference pilot tones across multiple symbols, the residual frequency shift can be calculated. RFO is a small frequency error relative to FFO derived using the PSS-CP correlations.

5.3   Frequency Correction in the Signal Path

Frequency error or correction value is applied at multiple stages distributed into RF synthesizers/LO and digital signal processing:

- As discussed in chapter 4, a big part of known correction can be applied to the crystal oscillator by varying the capacitors. As an example, a crystal oscillator starting at, say, 10 ppm can be set with different capacitor values that can regulate the adjusted ppm to less than 1–5 ppm.
- The next level correction goes to fractional-N frequency synthesizers, which are described in chapter 4. As given in chapter 4, synthesizers are capable of correcting up to a few hertz resolutions out of LTE UL/DL center frequencies.
- Synthesizers may usually take 30–100 μs to settle. In some designs, instead of updating the small residues to synthesizers, a digital demodulation on in-phase and quadrature-phase (I-Q) digital samples can be used, and this approach works as a frequency correction [URL (limemicro1)].
- As an extended digital processing scheme, fractional sampling Farrow filters are used to alter the sampling frequency of digital I-Q samples.

5.4   Carrier Frequency Offset Error and SNR Loss

In the previous section, Doppler influences are illustrated. In addition to Doppler, the frequency drifts can result from multiple sources, namely Rx chain local oscillator (LO) used with down-conversion, clock drifts at sampling, frequency error estimation, and any errors with digital samples frequency correction blocks. In this section, the frequency errors f_ε normalized to LTE OFDM tone resolution 15,000 Hz is referred to as ε (= $f_\varepsilon/15,000$) to make it consistent with that in [URL (NASA)]. The input signal power in one resource element (RE) is taken as E_s, and the noise plus interference as I_{ot} [URL (Slideshare)]—normalized to 15 kHz BW, same as resource element BW. The SNR loss increases with input E_s/Iot and ε as given in [URL (NASA)]. In the literature [URL (NASA)] E_c/N_0 is used, and for the LTE system E_s/I_{ot} is used as E_s/I_{ot} is more consistent with the LTE references [3GPP 36.133, URL (Slideshare)]. With a reference input SNR of E_s/I_{ot}, the modified SNR due to frequency offset is given as

$$\text{SNR\_ratio} \geq \frac{E_s}{I_{ot}}\left(\frac{\sin \pi\varepsilon}{\pi\varepsilon}\right)^2 \frac{1}{1+0.5947\left(E_s/I_{ot}\right)(\sin \pi\varepsilon)^2}. \qquad (5.1)$$

The E_s/I_{ot} is a power ratio, and if it is available in dB scale, use $E_s/I_{ot} =$

$10^{\wedge}[(E_s/I_{ot}$ in dB)$/10]$, and final calculation of SNR_ratio is a power ratio. To represent the SNR of equation (5.1) in dB, use $10 \log_{10}(\text{SNR\_Ratio})$. Figure 5.3 shows SNR loss for multiple E_s/I_{ot}. The x axis frequency error ε is normalized to 15 kHz, and the y axis is in dB scale.

Figure 5.3(a) is the amplitude loss with frequency offset. It is a plot of $\sin(x)/x$ where $x = \pi f_\varepsilon/15,000$, and f_ε is the frequency offset in hertz that could be contributed from Doppler, LO offset or any other drift due to sampling.

Figure 5.3(b) is the SNR degradation as a function of frequency offset for multiple input signal-to-noise plus interference ratios. The higher the input signal-to-noise ratio, the higher the SNR loss with frequency error.

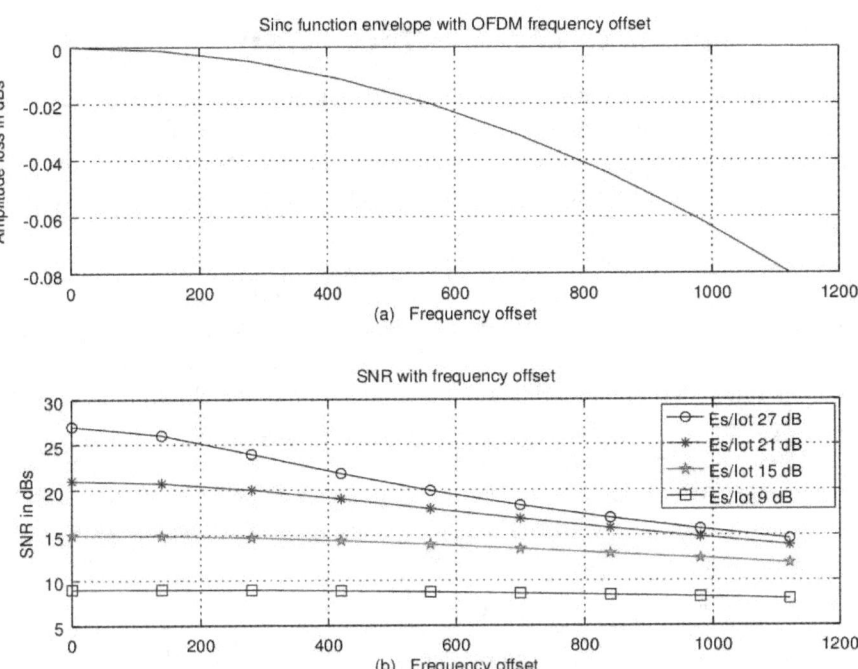

Figure 5.3. (a) Sinc function loss as a function of frequency offset for LTE OFDM. (b) SNR degradation as a function of frequency offset for multiple input signal-to-noise plus interference ratios.

Notes: From the SNR equation (5.1), it is not completely clear how SNR is influenced on the basis of LTE low BW 1.4 MHz signals and higher BW 20 MHz, and also the SNR loss deviations with QPSK, 16-QAM, and 64-QAM. This requires further investigation. In the LTE system, as the frequency error is made to settle within ±(0.1 ppm + 15 Hz), for Band 7 of carrier frequency 2,655 MHz, the frequency error is ±(265 +15 Hz). From numerical data used in the plots, the loss is approximately 3 dB for higher input SNRs of 27 dB, and 0.075 dB for 9 dB input SNR. On the basis of this formulation, the summary is to achieve better than ±(0.1 ppm + 15 Hz) frequency error for 64-QAM downlink signals. Further investigation is useful on SNR degradation with tone index. As a general indication, toward the band edges, SNR degradation may be higher. The SNR in equation (5.1) assumes that the band edges will have the highest loss. For the LTE tones that are not completely filled with the occupied BW, the SNR degradation can be lower than in equation (5.1) by (highest tone index)/N, where N is points in FFT. For LTE 20 MHz the highest tone index gives 1,200/2,048, and that may be the reason equation (5.1) uses SNR_ratio ≥ formulation math.

5.5   Sampling Clock Offset (SCO)

Sampling clock offset can be treated as frequency-domain or time-domain error. As described in previous sections, the carrier frequency offset between local oscillator reference and Doppler is required to be matched and tracked for improving signal quality. The SCO is the ADC and digital-to-analog converter (DAC) sampling clock error. In chapter 4, clock jitter and influence on ADC sampling and SNR loss are presented. The SCO is the frequency error, when sampling clock is not getting an exact number of samples in the one LTE symbol duration of $66\frac{2}{3}$ μs. The SCO error degrades SNR through signal loss and intercarrier interference (ICI). Taking the example of LTE 20 MHz BW, the basic sampling clock is 30.72 MHz or integer multiples of 30.72 MHz in case of oversampling.

Refer to [URL (entegro), URL (nctu)] on various SCO estimation approaches, which are not covered in this book. The SCO results in SNR loss mainly due to ICI observed at FFT output. The degradation loss is given as

$$D_n = 10 \log_{10}(1 + \frac{1}{3}\frac{E_s}{I_{ot}}(\pi n\, 10^{-6}\delta f_{sco})^2). \quad (5.2)$$

Here E_s/I_{ot} is a signal-to-noise power ratio as described with equation (5.1), δf_{sco} (here δf_{sco} is one single variable) is the SCO frequency error in ppm, the numerical value of δf_{sco} for 5 ppm is 5, and the factor 10^{-6} is for ppm (1/million). The degradation or loss is tone dependent, and loss increases at the band edge. Table 5.2 lists various ppm and loss combinations. (Note: To keep the representation compact ± is not included with the numerical values.)

To translate the issues into practical implementation, the ADC/DAC clocks are generated from a base reference of either crystal oscillator (XO) or TCXO. A good quality TCXO maintains 0.25–0.5 ppm. On the basis of table 5.2, the loss is not significant even if small ppm is left uncorrected. The TCXO frequency may not be directly compatible for the ADC oversampling; hence, a phase locked loop (PLL) or a fractional-N synthesizer is commonly used with the divider.

To minimize losses with TCXO scheme or to account for higher ppm TCXOs, a Farrow filter may be used. Farrow filter alters the sampling rate with finer resolution. The Farrow filters are commonly used on oversampled ADC output.

Table 5.2. SCO loss listed for various ppm errors and input SNR combinations

SCO ±ppm	Input SNR (Es/Iot) in dB (9 to 40 dB)					
	40	36	27	21	15	9
0.1	0.00051	0.00020	0.00003	0.00001	0.00000	0.00000
0.25	0.00321	0.00128	0.00016	0.00004	0.00001	0.00000
0.5	0.01283	0.00511	0.00064	0.00016	0.00004	0.00001
0.56	**0.01608**	**0.00641**	**0.00081**	**0.00020**	**0.00005**	**0.00001**
1	0.05108	0.02041	0.00257	0.00065	0.00016	0.00004
2	0.20082	0.08106	0.01029	0.00259	0.00065	0.00016
3	0.43946	0.18031	0.02312	0.00582	0.00146	0.00037
4	0.75294	0.31556	0.04101	0.01034	0.00260	0.00065
5	1.12535	0.48347	0.06391	0.01614	0.00406	0.00102
10	3.39087	1.67621	0.25019	0.06421	0.01622	0.00408
15	5.63734	3.13828	0.54392	0.14316	0.03641	0.00917
0.56 ppm is UL reference with HS train (0.46 ppm at 500 km/h) and 0.1 ppm UE maximum error						

When frac-N synthesizers with good resolution is used with TCXO, the SCO errors can be programmed to frac-N synthesizer, and ADC sampling clock can be tuned to the required precision levels.

In the literature, examples are shown with crystal oscillator (XO), and XO is of higher ppm thus creating more SNR loss through ICI. Use of Farrow filter is the digital signal processing approach, but for XO-based approach, the better option would be to use a frac-N synthesizer and correct the errors to the required value, and any residual errors may be taken care of with Farrow filters or left as uncorrected on the basis of the allowed degradation with reference to table 5.2.

Table 5.2 is marked with additional clarification notes for 0.56 ppm. When UE is working with eNodeB, the UL frequency is maintained to within ±0.1 ppm of reception frequency. The reception at UE can be with maximum deviation of ±0.46 ppm for a high-speed train at 500 km/h. By using simply UL carrier reference as SCO, the maximum error is ± 0.56 ppm, and the corresponding losses are marked in table 5.2 in the row marked with 0.56 ppm, and the losses or degradations at 0.56 ppm are not significant for UE applications. Table 5.3 lists SCO correction approaches and the relative merits.

Table 5.3. Sampling clock offset correction approaches

XO, and TCXO architectural options with synthesizers and Farrow filters	Description of the selected option and trade-offs
TCXO → synthesizer without SCO correction → sampling clock divider	Simplest and can be used in low throughput UE devices up to 16-QAM based on selected TCXO ppm. Such devices may not be relevant for high Doppler from 500 km/h high speed train. Some TCXO with 0.25 ppm may allow direct usage.
TCXO → Int-N PLL → sampling clock divider → correct error in Farrow filters	This is to implement using Farrow filters in signal processing. TCXO reference is directly used with sampling clock and Farrow filters can keep track of SCO updates.
TCXO → frac-N synthesizer → sampling clock divider	Simpler and nice implementation and can be tuned to frac-N higher resolution—when SCO estimates are available. This is to generate precise sampling clock instead of pushing the problem to Farrow filters in digital signal processing.
XO → frac-N synthesizer → sampling clock divider →	Simpler and frac-N higher resolution can absorb all the other errors and SCO estimates.
XO → frac-N synthesizer → sampling clock divider → Farrow filters	frac-N synthesizer can take care of absorbing all the errors. In case synthesizer is held steady; short term variations can be incorporated in to Farrow filters.

5.6   Establishing Reference Quality with Frequency Errors

The following summary lists establishing the reference quality by isolating frequency errors:

- In a direct connection between eNodeB (in the figures instruments replicate the eNodeB) and UE, the instrument is directly connected to the UE through RF components, such as attenuators, splitters, filters, duplexers, and circulators. The input at UE is Doppler free, and the channel is stationary. In the configuration shown in figure 5.4(a), the frequency errors in UE are decided by the crystal/TCXO reference frequency ppm error, synthesizer frequency errors, and any residual estimation errors in the algorithms.

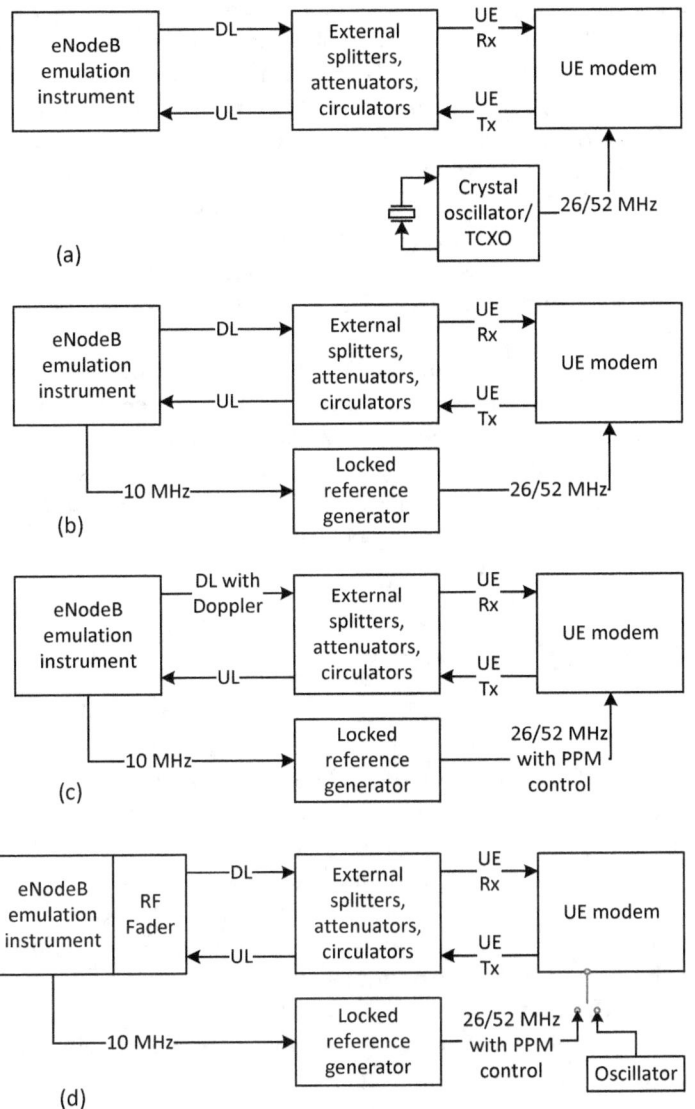

Figure 5.4. (a) Tx and Rx with free running oscillator. (b) Rx with frequency locked reference. (c) Rx with Doppler and frequency locked reference. (d) With fader that can create several impairments.

- To isolate these influences, as indicated in figure 5.4(b), the reference signal at 10 MHz from the eNodeB instruments is fed to the UE. If the UE is not capable of taking 10 MHz reference, an optional frequency source can be used that locks to eNodeB reference and provides UE compatible frequency reference. As an example UE needs 26 MHz, the function of frequency locking source is to take 10 MHz reference from eNodeB instrument and produce 26 MHz output. With this approach the UE reference will be tracking the eNodeB reference frequency. The remaining residual errors are related to synthesizers used for LOs and from mixed-signal clocks. If there are any systematic or deterministic errors in the synthesizers, they may be minimized through various configurations.

- With the listed previous conditions, the performance of receiver and transmitter is supposed to be best achievable from frequency errors perspective.
- The block diagram configuration shown in figure 5.4(c) introduces the Doppler, and in another operation, the locked reference can also be drifted by known ppm to characterize the UE in a controlled way.
- The configuration shown in figure 5.4(d) introduces the fader that could be part of the same instrument to simulate several conditions. Again the clock reference can be from a free running oscillator or precisely ppm controlled locked reference.

5.7   Conclusive Remarks

Frequency errors can have significant influence on the signal quality. The issues have to be resolved by architecture and design. In LTE design when the product has affordability in bill of material (BOM), use better ppm reference such as TCXO that eliminates several issues. Factory calibration marked in this chapter helps UE come up with small frequency offsets. The SCO when taken care of to small residual ppm, the losses are minimal. Overall the frequency offsets or sampling clock offsets cannot be eliminated to zero. The goal is to minimize these frequency residues to the extent that the degradation contributions stay in a fraction of a dB—of the order of 0.1–0.2 dB. Any of the corrections through hardware and firmware in a closed loop has to maintain stable (without overshoots) operation with acceptable drift or frequency wandering.

6 RECEIVER I-Q IMPAIRMENTS

In the receiver, in-phase and quadrature-phase (IQ or I-Q) impairments, usually referred to as I-Q imbalances or distortions start from the receiver mixer stage onward and ends at the analog-to-digital converter (ADC) output [URL (Keysight-John)]. Refer to [URL (Charon), URL (Mathworks-IQ)] for short introduction on this topic. Figure 6.1 is marked with various blocks of the receiver chain to convey imbalance contributions. Consider a reception of 1 tone f_{RF} at radio frequency (RF) represented as $f_{RF} = f_{lo} + f$ at mixer input. Here f_{lo} is the down-conversion local oscillator (LO) frequency, and f is the baseband signal frequency. The mixer input is fed through front-end electronics and low-noise amplifier (LNA). The mixer in LTE system receiver is a zero intermediate frequency (IF) down-converter. The low pass filtered I-Q signals from mixer output at frequency f is denoted as follows [Sivannarayana, 1998]:

$$x(t) = A\,e^{j\omega t} = A\,[\cos(\omega t) + j\sin(\omega t)] = A\cos(\omega t) + j\,A\sin(\omega t) \quad (6.1)$$

where $x(t)$ is the filtered I-Q output analog signals, A is the amplitude, $\omega = 2\pi f$, $A\cos(\omega t)$ is the real or I-signal, and the term with operator j, $A\sin(\omega t)$ is the imaginary part or Q-signal. Here j is the imaginary term operator (i.e., sqrt(−1)). The I-Q signals can undergo amplitude, phase, and DC offset differences. The local oscillator (LO) used in down-conversion uses two LO components with 0 and 90 degrees shifted for I and Q generation, and these LO signals can be of different in amplitude and deviate in phase from 90 degrees. The LO phase differences in I and Q paths can create a relative phase shift between I and Q signals. The mixer characteristics for I and Q paths could be different, contributing to amplitude difference. The impedance mismatches and various LO and RF leakages create a DC (direct current or voltage) offset. The amplitude and phase mismatches or imbalances in I and Q are relative to one another, and they are usually shown in one of the signals with reference to another. In the following representation, amplitude imbalance term $(1 + \varepsilon)$ is shown in I-signal relative to Q, and phase imbalance θ in Q-path relative to I:

$$x(t) = A\,[(1 + \varepsilon)\cos(\omega t) + j\sin(\omega t + \theta)]. \quad (6.2)$$

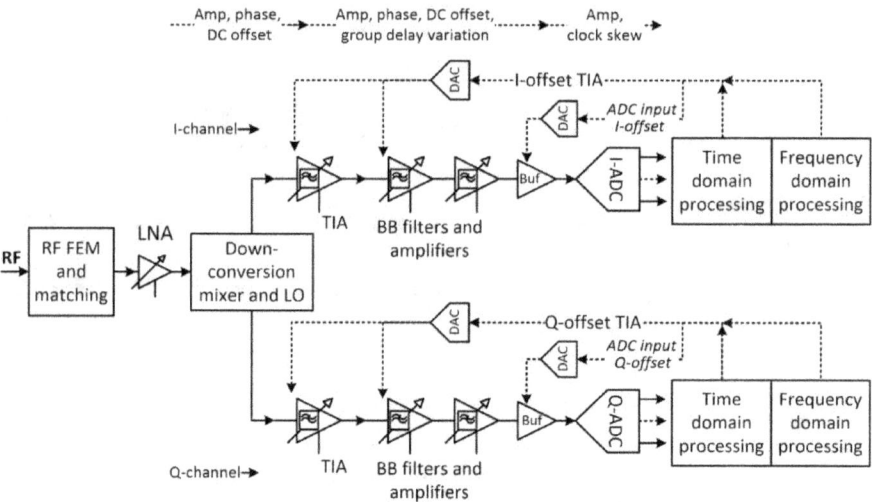

Figure 6.1. I-Q imbalances in a receiver.

If the mixer is of current source, the mixer outputs may go through transimpedance amplifier (TIA) that creates current to voltage conversion; refer to [URL (jos)] for an extended overview on this topic of mixer architectures with TIA. To suppress unwanted components, TIA incorporates low pass filtering operation. The TIA may introduce additional mismatches of amplitude (gain difference between I-Q paths), phase difference due to difference in filters response, and direct current (DC) offsets.

The TIA output is further amplified by 30–45 dB in analog amplifier/filter sections to provide sufficient gain before sampling at ADCs. The analog sections gain is achieved with two or three stages of amplification. The mixer output is of wide-band, and low pass filters are required to suppress unwanted blocker signals, interferences, and wide-band noise. The amplifiers in baseband also incorporate antialiasing filtering operation for ADCs. Multi-stage amplifiers can introduce amplitude imbalance (due to gain difference), and filter component deviations can introduce phase imbalance. The group delay is the derivative of phase difference. The filters in I-Q path may introduce variation in phase response usually apportioned as group delay variation. The difference in group delay between I-Q paths is a frequency-selective (dependent) phase imbalance. Added to all these, high-gain 30–45 dB amplifiers used as cascaded filters introduce different DC offset among I and Q signals.

The baseband amplifiers drive I and Q ADCs. The ADC can have amplitude mismatch, meaning ADC-I will have full scale of, say, 1 V (volt) and ADC-Q will have full scale of 0.95 V. The ADC clock applied to I and

Q may be buffered differently and can introduce clock skew, or time delay. With time delay, the analog signal in Q will be captured at a different time instant relative to I. The Q-sample capture time instant will either lead or lag in time with reference to I. As explained in later sections, a fixed time delay between I-Q introduces linear phase shift with frequency, resulting in frequency-dependent phase imbalance.

Once I-Q signals cross ADC, there are no new I-Q imbalances, and the main follow-on operations specific to I-Q signals are imbalance estimation and helping to postcancel imbalances created in RF, analog, and mixed-signal sections.

Assuming digital hardware and firmware are able to estimate the imbalance parameters, the DC offset can be fed back to TIA or first stages of baseband amplifier/filters that removes the coarse-level DC offset. Note the DC offset is independent in I and Q paths.

The residual offset (after subtracting coarse-level offset in analog sections) accumulated through baseband stages cross the ADC and can be cancelled in digital. If the DC residual offset is significant at ADC input, a next level fine offset correction can be included at the ADC input buffer stages as shown in figure 6.1.

Out of five imbalances/impairments described so far, the three items—amplitude, phase, and DC offset—are primarily approximated as frequency independent, and I-Q sampling time delay-offset and group delay variation are frequency dependent.

6.1   I-Q Imbalance Illustrations

Figure 6.2 illustrates imbalances. Refer to [URL (R&S-1)] for some more material on this topic. Figure 6.2(a) is for UE1 that has tones on the right side of zeroth-indexed tone and imbalances appear on the left side. The x axis is the frequency, and the imbalance is a mirror on the y axis. The kth tone imbalance appears at $-k$ (on the x axis) indexed tone. This representation f_0 tone at k, and $-k$ appears as not a problem to process desired tones on the right side. The DC offset is always present as a zeroth tone, and its imbalance image appear at the same zeroth bin. Figure 6.2(b) is for the UE2 tones on the left side and imbalance appearing on the right side. Starting with simple assumption that eNodeB is serving just UE1 and UE2 only then UE1 receives both UE1 and UE2 tones. Similarly UE2 receives UE2 and UE1 tones. UE2 tones are unwanted in UE1, but UE1 processes these until frequency domain. As shown in figure 6.2(c) the UE2 tones create imbalance for UE1.

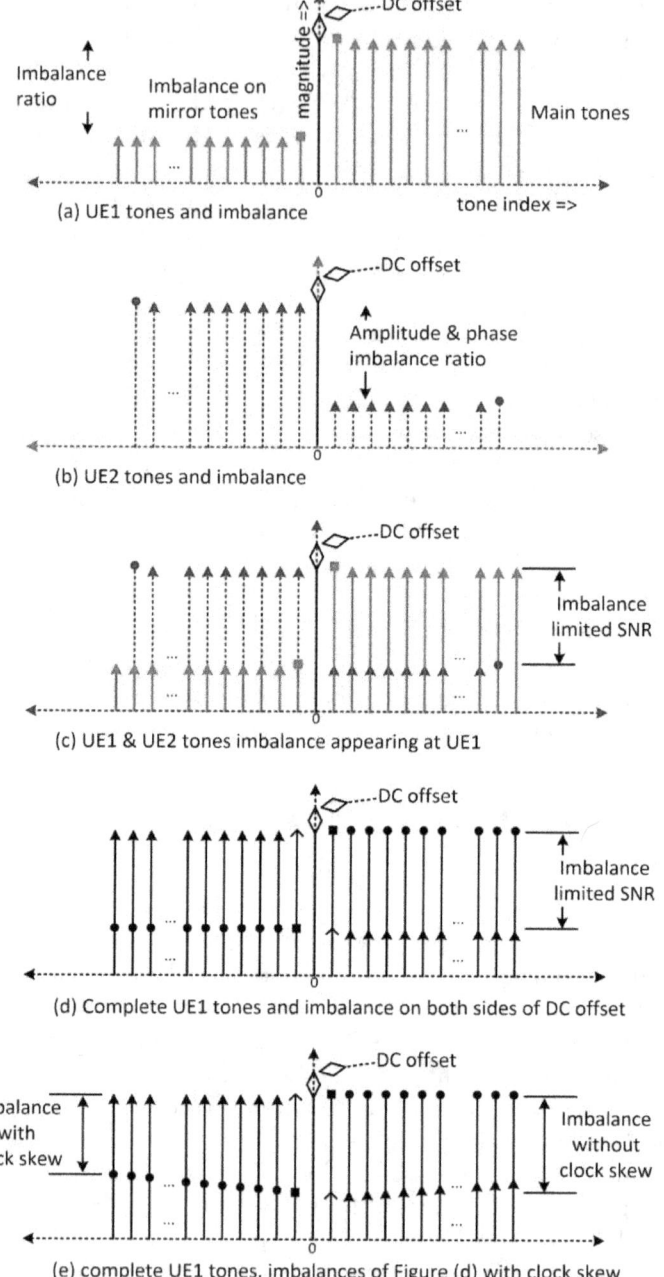

Figure 6.2. (a) to (e) I-Q imbalance illustrations [x axis is tone index and y axis is signal level, also marked for imbalance levels].

Even if noise limited signal-to-noise ratio (SNR) is good, the imbalance can increase the floor and limit the SNR. Figure 6.2(d) shows eNodeB is fully providing (serving) all tones to UE1 only. In this case the imbalance from left side tones falls on right and vice versa from the same UE1. The tones and the corresponding matched imbalance pairs are marked with different arrowheads for clear visual distinction. Figure 6.2(e) is for the I-Q imbalance along with the clock skew. The clock skew generates growing imbalance with frequency. There will be fixed imbalance floor built up due to amplitude and phase and a frequency-dependent imbalance (ramp) builds up on the floor. This causes band-edge tones to suffer more. As a continuity of these illustrations, mathematical expressions for various imbalances are presented in next sections.

The following representation considers the amplitude $(1 + \varepsilon)$, phase θ, DC offset in I (as Ioff) and DC offset in Q (as Qoff) and time delay/skew δt. The impairments at frequency f can be represented together as follows.

$$x(t) = A\,[\,\{\text{Ioff} + (1 + \varepsilon)\cos(\omega t)\} + j\,\{\text{Qoff} + \sin(\omega(t + \delta t) + \theta)\}\,]. \quad (6.3)$$

Notes: There are many valid variations to this representation depending on how the errors are accounted among I and Q relative to one another. In the next chapter, transmitter I-Q impairments are presented and followed by a few basic steps on resolving—estimation and cancellation—of these impairments.

6.2   DC Offset Influencing the Signal Dynamic Range

DC offsets without cancellation can limit the dynamic range of the baseband analog amplifiers/filters and ADCs. A DC offset of 0.1 V propagated to amplifier designed with good linear region in 1 V peak-peak (pk-pk) limits the usage of amplifier to 0.8 V pk-pk. The ADC with full scale of 1.0 V pk-pk with amplified offset of 0.2 V limits the ADC usage for AC signal swing of 0.6 V pk-pk, a loss of $20\log_{10}(0.6/1.0) = -4.44$ dB. By adding DC offset cancellation after RF mixer output stage or at subsequent amplifier stages reduce the offsets at ADC input. The offset cancellation in analog sections can be at coarse level and the residue can be made to stay at approximately well within ± 50 to ± 100 mV at ADC input. For ± 50 mV a loss of $20\log_{10}(0.9/1.0) = -0.92$ dB is generally acceptable. In figure 6.1, a DC offset cancellation is also shown at the input of ADC. In case the baseband amplifiers are introducing higher offsets, a second level of DC offset cancellation may be desired. The time-domain and frequency-domain digital processing modules can feedback the DC offset estimates to low resolution independent I-Q DACs to reduce DC offsets. The DC offsets are sensitive as the LTE device scans within the intraband (intra) and interband. The DC offset minimization has to be an adaptive approach as it has dependency on gain, BW, and frequency band.

> Notes: The ADCs in the current technology may operate on hundreds of mV of analog signal input. For easy in conveying examples 1 V pk-pk is used in this book.

Baseband filters for minimizing the imbalances. LTE has six BWs from 1.4 to 20 MHz, and baseband BW for 20 MHz LTE is 9 MHz. The baseband analog filters are made wider than 9 MHz (in 20 MHz mode) to reduce group delay (in turn minimize frequency-dependent phase imbalance). The useful signal is maintained in the flat filter region and 9 MHz frequency response stays within approximately 0.5–1 dB from flat response.

6.3   Modeling the I-Q Imbalances

I-Q channels are modeled in four or more ways depending on the presence of the error terms either in I or Q. The representation in [Sivannarayana, 1998] is used here.

$$x(t) = a(t)[(1 + \varepsilon)\cos(\omega t) + j\sin(\omega t + \theta)]. \qquad (6.4)$$

where $a(t)$ is the amplitude envelope, and ω is the angular frequency in radians, same as $2\pi f$ with f being frequency in hertz, ε is the relative amplitude error between I-Q, θ is the phase error in radians (not degrees) between I-Q signals and t is the time in seconds.

I-Q signals without any errors match in amplitude, phases, and maintain zero DC offsets. Figure 6.3(a) is with no errors that give clear circle $\sqrt{\sin^2(\omega t) + \cos^2(\omega t)} = 1$ with magnitude (radius) always unity. As an example, the QPSK constellation points are marked in circles that stay at the desired location. For easy identification of start and end points, the plots are shown with less than one full cycle.

When I-Q signals are subjected to errors, the radius of the circle is not unity. Referring to figure 6.4(a), the I-cosine wave is reduced in amplitude by 20%, and the amplitude or magnitude along the x axis (I-amplitude) is lower as given by $\sqrt{[0.8\cos(\omega t)]^2 + [\sin(\omega t)]^2}$. The QPSK ideal constellation points are marked in circles, and the imbalanced signal constellation points are marked in * that deviates from the desired location.

Figure 6.4(b) includes only phase error that tilts the circle, and the magnitude deviates from unity given as $\sqrt{[\cos(\omega t)]^2 + [\sin(\omega t + \theta)]^2}$.

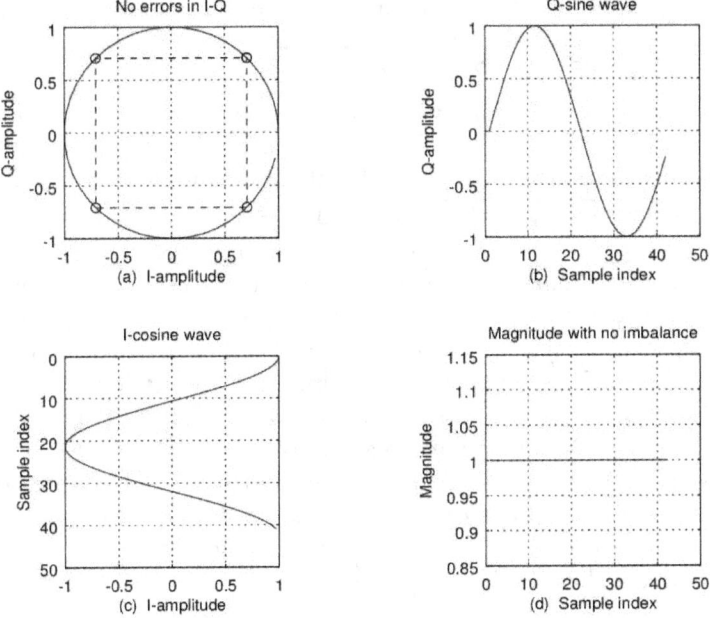

Figure 6.3. I-Q single tone signal without any errors.

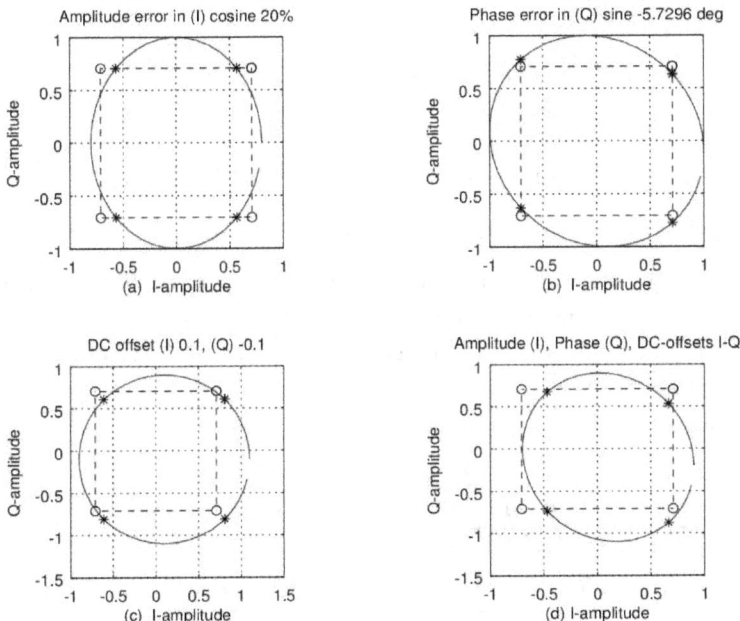

Figure 6.4. I-Q single tone constellation with errors; the marking o is desired, and * is with imbalances.

Figure 6.4(c) includes only DC offsets that move the magnitude in x and y coordinates, and creates magnitude deviation from unity $\sqrt{[0.1 + \cos(\omega t)]^2 + [-0.1 + \sin(\omega t)]^2}$.

Figure 6.4 (d) is for the combined errors of amplitude, phase and DC offsets. The QPSK ideal constellation points are marked in circles, and the imbalanced points are marked in * that deviate from the desired location given as $\sqrt{[0.1 + 0.8 \cos(\omega t)]^2 + [-0.1 + \sin(\omega t + \theta)]^2}$.

On the basis of the errors, the constellation points appear with deviations on x-y coordinates, and in addition they also create image (mirror) frequency components. For easy identification of start and end points, the plots are shown with less than one full cycle.

Note that the shift of constellation points shown in figures 6.4(a) through (d) is not a problem for QPSK signals. In fact 16-QAM (quadrature amplitude modulation) may also tolerate this level of imbalances. When 64-QAM is used the imbalances have to be corrected before taking into frequency-domain analysis of OFDM signal. The deviations in constellation points results in reduction of error vector magnitude (EVM), as described in chapter 10.

6.4   I-Q Imbalance Mathematical Formulation

The amplitude and phase imbalances are represented as follows. For simplicity amplitude envelope $a(t)$ is taken as A.

$$x(t) = A\left[(1 + \varepsilon) \cos(\omega t) + j \sin(\omega t + \theta)\right]$$

$$x(t) = {}^A\!/_2 \left[(1 + \varepsilon)\left(e^{j\omega t} + e^{-j\omega t}\right) + \left(e^{j(\omega t + \theta)} - e^{-j(\omega t + \theta)}\right) \right]$$

$$x(t) = {}^A\!/_2 \left[e^{j\omega t}\left(1 + \varepsilon + e^{j\theta}\right) + e^{-j\omega t}\left(1 + \varepsilon - e^{-j\theta}\right) \right].$$

With imbalances, a new term is appearing $e^{-j\omega t}$ that represents the image frequency. The ratio of image frequency strength $X(-\omega)$ associated with $e^{-j\omega t}$ to the main component $X(\omega)$ that is associated with $e^{j\omega t}$ is a complex number denoted as follows:

$$\frac{X(-\omega)}{X(\omega)} = \frac{(1 + \varepsilon - e^{-j\theta})}{(1 + \varepsilon + e^{j\theta})}$$

$$\frac{X(-\omega)}{X(\omega)} = \frac{[1 + \varepsilon - \cos(\theta) + j \sin(\theta)]}{[1 + \varepsilon + \cos(\theta) + j \sin(\theta)]}$$

$$\frac{X(-\omega)}{X(\omega)} = \frac{[\varepsilon^2 + 2\varepsilon + j\, 2(1 + \varepsilon)\, \sin(\theta)]}{[2(1 + \varepsilon + \frac{\varepsilon^2}{2} + \varepsilon \cos(\theta) + \cos(\theta)]} \approx \frac{\varepsilon}{2} + j\frac{\theta}{2}$$

For small values of ε and θ the voltage ratio $X(-\omega)/X(\omega)$ and power ratio

$P(-\omega)/\ P(\omega)$ can be approximated to

$$\frac{X(-\omega)}{X(\omega)} \approx \frac{\varepsilon}{2} + j\frac{\theta}{2} \ and \ \frac{P(-\omega)}{P(\omega)} \approx \frac{\varepsilon^2}{4} + \frac{\theta^2}{4}. \quad (6.5)$$

The voltage ratio and power ratio for four different I-Q signal imbalance models are given in table 6.1. The mathematical derivations for all the 4-combinations are not shown, but the results are given in table 6.1. The voltage ratio with proper sign has to be used for imbalance correction as power ratio keeps only magnitude.

Table 6.1. I-Q signal model and imbalance ratios [Sivannarayana,1998]

Signal model x(t)	$X(-\omega)/X(\omega)$	$P(-\omega)/\ P(\omega)$
$A\,[(1 + \varepsilon)\cos(\omega t) + j\sin(\omega t + \theta)]$	$(\varepsilon/2 + j\ \theta/2)$	$\varepsilon^2/4 + \theta^2/4$
$A\,[\cos(\omega t) + j\,(1 + \varepsilon)\sin(\omega t + \theta)]$	$(-\varepsilon/2 + j\ \theta/2)$	$\varepsilon^2/4 + \theta^2/4$
$A\,[\cos(\omega t + \theta) + j\,(1 + \varepsilon)\sin(\omega t)]$	$(-\varepsilon/2 - j\ \theta/2)$	$\varepsilon^2/4 + \theta^2/4$
$A\,[(1 + \varepsilon)\cos(\omega t + \theta) + j\sin(\omega t)]$	$(\varepsilon/2 - j\ \theta/2)$	$\varepsilon^2/4 + \theta^2/4$

The essential point from table 6.1 is to use suitable signal model that matches the adapted imbalance estimation and calibration, and use imbalance correction with proper sign.

Notes on amplitude and gain imbalance usage: The amplitude imbalance in this book is the signal level difference between I and Q signals. The signal level imbalance may be created at multiple stages, and amplifier gain difference in I and Q creates amplitude imbalance that can be referred as gain imbalance. Referring to amplifier stage amplitude imbalance, often gain imbalance name is used. In general these two names amplitude and gain imbalance are used interchangeably, and such usage is accepted in the industry and literature.

6.5   I-Q Imbalance Numerical Examples

Table 6.2 lists the phase imbalance arithmetic. The data sheets usually list θ in degrees. This has to be converted to radians, so recall that θ in radians = (θ in degrees) $(\pi/180)$. In this chapter and next chapter unless it is stated otherwise, θ is assumed to be in radians. The imbalance ratio is $20 \log_{10}(\theta/2)$ dB, where θ is in radians. A phase imbalance change by a factor of 2 creates phase imbalance change of 6 dB.

Table 6.2. Phase imbalance as per approximations

Phase imbalance arithmetic			
Phase error (deg)	Phase error (radians)	Phase imbalance in dB	Notes on phase imbalance
0.6	0.010	−45.619	
0.8	0.014	−43.120	
1.0	0.017	−41.183	
1.2	0.021	−39.598	Good level to target for LTE 64-QAM
1.4	0.024	−38.259	
1.6	0.028	−37.100	
1.8	0.031	−36.076	
2.0	0.035	−35.161	
2.2	0.038	−34.333	
2.4	0.042	−33.578	Acceptable in some designs
2.6	0.045	−32.882	
2.8	0.049	−32.239	
3.0	0.052	−31.640	
3.2	0.056	−31.079	
3.4	0.059	−30.552	
3.6	0.063	−30.056	
3.8	0.066	−29.586	
4.0	0.070	−29.141	
4.2	0.073	−28.717	
4.4	0.077	−28.313	
4.6	0.080	−27.927	
5.0	0.087	−27.204	
6.0	0.105	−25.620	
7.0	0.122	−24.281	
8.0	0.140	−23.121	
9.0	0.157	−22.098	
10.0	0.175	−21.183	

The amplitude imbalance error ε is a factor. The data sheet usually lists amplitude error in $\pm$dB that corresponds to $20 \log_{10}(1+\varepsilon)$ and $20 \log_{10}(1-\varepsilon)$. Table 6.3 lists various forms of entries to map from one to another. The amplitude imbalance is $20 \log_{10}(\varepsilon/2)$. Figure 6.5 presents I-Q amplitude imbalance $20 \log_{10}(\varepsilon/2)$ dBs versus percentage error. Refer to figure 6.5 while looking at combined amplitude and phase imbalances in tables 6.2 and 6.3.

Table 6.3. Amplitude imbalance approximations

Amplitude/gain imbalance arithmetic							
Gain error ε%	Gain error factor ε	Gain imbalance (1+ε) factor	Gain error 20 $\log_{10}$ (1+ε) dB	Gain imbalance (1−ε) factor	Gain error 20 $\log_{10}$ (1−ε) dB	I-Q gain imbalance 20 $\log_{10}$ (ε/2) dB	Notes on amplitude imbalance
1.0%	0.010	1.010	0.09	0.990	−0.09	−46.021	
2.0%	0.020	1.020	0.17	0.980	−0.18	−40.000	Good level to target
2.5%	0.025	1.025	0.21	0.975	−0.22	−38.062	
3.0%	0.030	1.030	0.26	0.970	−0.26	−36.478	
3.5%	0.035	1.035	0.30	0.965	−0.31	−35.139	
4.0%	0.040	1.040	0.34	0.960	−0.35	−33.979	Acceptable in some designs
4.5%	0.045	1.045	0.38	0.955	−0.40	−32.956	
5.0%	0.050	1.050	0.42	0.950	−0.45	−32.041	
6.0%	0.060	1.060	0.51	0.940	−0.54	−30.458	
7.0%	0.070	1.070	0.59	0.930	−0.63	−29.119	
8.0%	0.080	1.080	0.67	0.920	−0.72	−27.959	
9.0%	0.090	1.090	0.75	0.910	−0.82	−26.936	
10.0%	0.100	1.100	0.83	0.900	−0.92	−26.021	
11.0%	0.110	1.110	0.91	0.890	−1.01	−25.193	
12.0%	0.120	1.120	0.98	0.880	−1.11	−24.437	
13.0%	0.130	1.130	1.06	0.870	−1.21	−23.742	Pre-calibration levels in multiband receivers
14.0%	0.140	1.140	1.14	0.860	−1.31	−23.098	
15.0%	0.150	1.150	1.21	0.850	−1.41	−22.499	
16.0%	0.160	1.160	1.29	0.840	−1.51	−21.938	
17.0%	0.170	1.170	1.36	0.830	−1.62	−21.412	
18.0%	0.180	1.180	1.44	0.820	−1.72	−20.915	
19.0%	0.190	1.190	1.51	0.810	−1.83	−20.446	
20.0%	0.200	1.200	1.58	0.800	−1.94	−20.000	

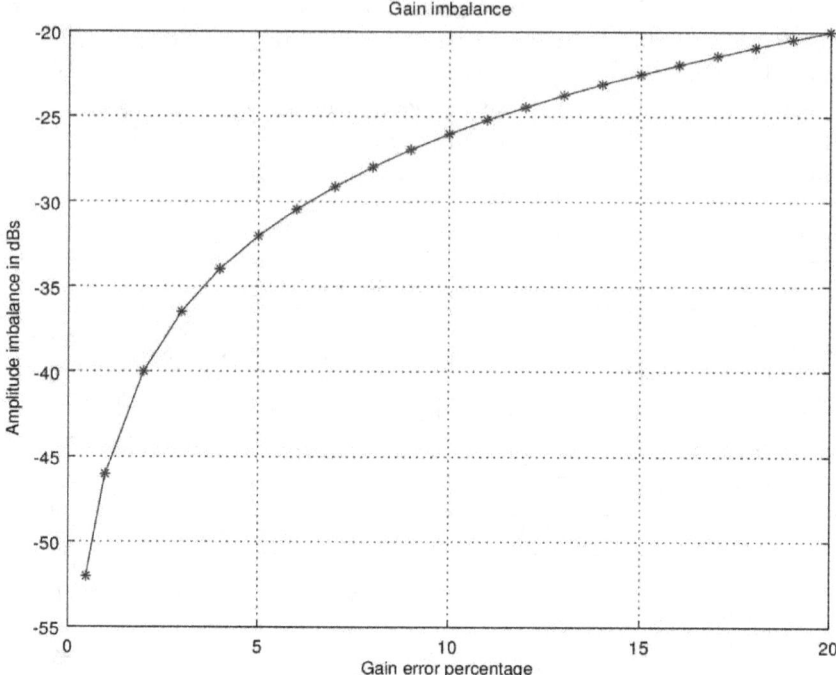

Figure 6.5. Gain (amplitude) imbalance with gain error percentage.

6.6   Trade-offs in Phase and Amplitude Balance

As shown in figure 6.6, improving entirely either amplitude (gain) balance or phase balance will not help. The goal is to arrive at a combined improvement. Figure 6.6 shows the combined response of amplitude and phase imbalance. The linear region is the most useful combined balance. Taking example of 1-degree phase imbalance, the phase imbalance is −41.18 dB. If gain error is 2%, amplitude imbalance is −40 dB. These two values are close and the combined imbalance is optimum. Suppose the system has limitation of 3% gain error that is −36.5 dB amplitude imbalances, the benefit out of 1-degree phase imbalance is masked by amplitude imbalance. The amplitude imbalance is dominating in this example.

Using ε and θ (in radians), $10 \log_{10}(\varepsilon^2/4 + \theta^2/4)$ provides direct combined imbalance. When amplitude and phase imbalances are available in dB scale, the combined imbalance can be obtained similar to section 3.6 and table 3.3.

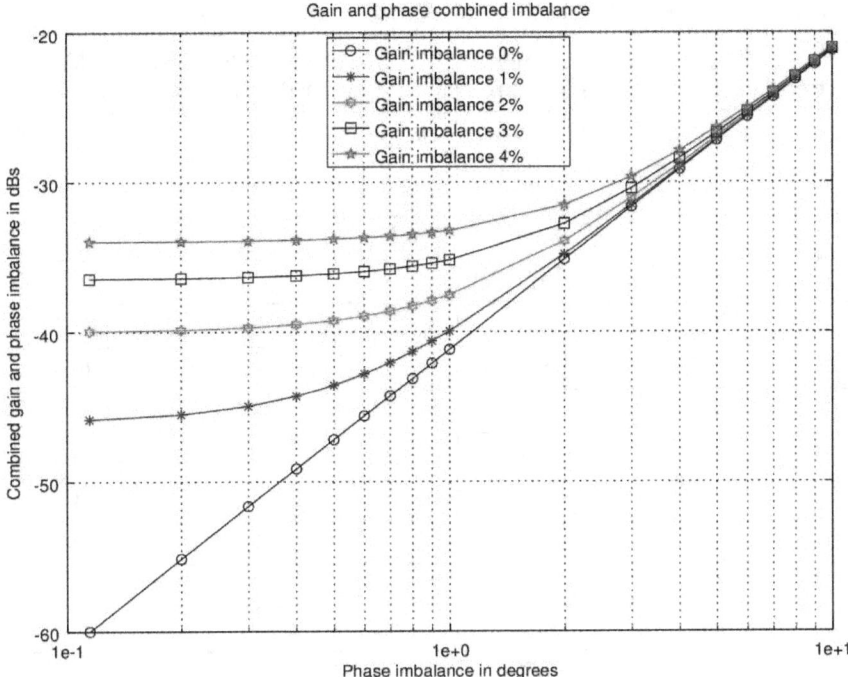

Figure 6.6. Gain and phase imbalance combined influence.

6.7   Time Delay or Clock-Skew Imbalance

The time-delay (often referred to as clock-skew) is relative time difference between I-Q components that introduces linear phase shift [URL (Delay1), URL (Keysight-John)]. LTE band-edge higher-frequency tones get higher phase imbalance (figure 6.7). In the following formulation, delay δt is introduced in Q-signal relative to the I-signal. Let

$$x(t) = A\left[\cos(\omega t) + j\sin(\omega(t + \delta t))\right] = A\left[\cos(\omega t) + j\sin(\omega t + \omega\delta t)\right]$$

$$x(t) = {}^{A}\!/_{2}\left[(e^{j\omega t} + e^{-j\omega t}) + (e^{j(\omega t + \omega\delta t)} - e^{-j(\omega t + \omega\delta t)})\right]$$

$$x(t) = {}^{A}\!/_{2}\left[e^{j\omega t}\left(1 + e^{j\omega\delta t}\right) + e^{-j\omega t}\left(1 - e^{-j\omega\delta t}\right)\right]$$

$$\frac{X(-\omega)}{X(\omega)} = \frac{(1-e^{-j\omega\delta t})}{(1+e^{j\omega\delta t})} = j\left[\frac{\sin(\omega\delta t)}{(1+\cos(\omega\delta t))}\right] \approx \frac{j\omega\delta t}{2} = \frac{j2\pi f\delta t}{2} = j\pi f\delta t.$$

In the above formulation as $\omega\delta t$ is a small quantity, $\sin(\omega\delta t) \approx \omega\delta t$ and $\cos(\omega\delta t) \approx 1$. In case of LTE, the frequency bin width (δf) is 15,000 Hz, and the imbalance for a frequency bin or tone index n is given as $\pi f\delta t = (\pi\, n\, \delta f\, \delta t) = (15{,}000\, \pi\, n\, \delta t)$. Here f in $\pi f\delta t$ is replaced with $n\, \delta f$, and then δf with 15,000 Hz.

Table 6.4. I-Q relative delay and phase imbalance

LTE 10 MHz channel BW ⇒ 4.5 MHz baseband signal delay imbalance			LTE 20 MHz channel BW ⇒ 9 MHz baseband signal delay imbalance		
Delay in ns	Band edge frequency in MHz	Delay specific phase imbalance in dB	Delay in ns	Band edge frequency in MHz	Delay specific phase imbalance in dB
0.1	4.5	−57.0	0.1	9.0	−51.0
0.2	4.5	−51.0	0.2	9.0	−45.0
0.3	4.5	−47.5	0.3	9.0	−41.4
0.4	4.5	−45.0	**0.4**	**9.0**	**−38.9**
0.5	4.5	−43.0	0.5	9.0	−37.0
0.6	4.5	−41.4	0.6	9.0	−35.4
0.7	4.5	−40.1	0.7	9.0	−34.1
0.8	**4.5**	**−38.9**	0.8	9.0	−32.9
0.9	4.5	−37.9	0.9	9.0	−31.9
1.0	4.5	−37.0	1.0	9.0	−31.0
1.1	4.5	−36.2	1.1	9.0	−30.1
1.2	4.5	−35.4	1.2	9.0	−29.4
1.3	4.5	−34.7	1.3	9.0	−28.7
1.4	4.5	−34.1	1.4	9.0	−28.0
1.5	4.5	−33.5	1.5	9.0	−27.5
1.6	4.5	−32.9	1.6	9.0	−26.9
1.7	4.5	−32.4	1.7	9.0	−26.4
1.8	4.5	−31.9	1.8	9.0	−25.9
1.9	4.5	−31.4	1.9	9.0	−25.4
2.0	4.5	−31.0	2.0	9.0	−25.0

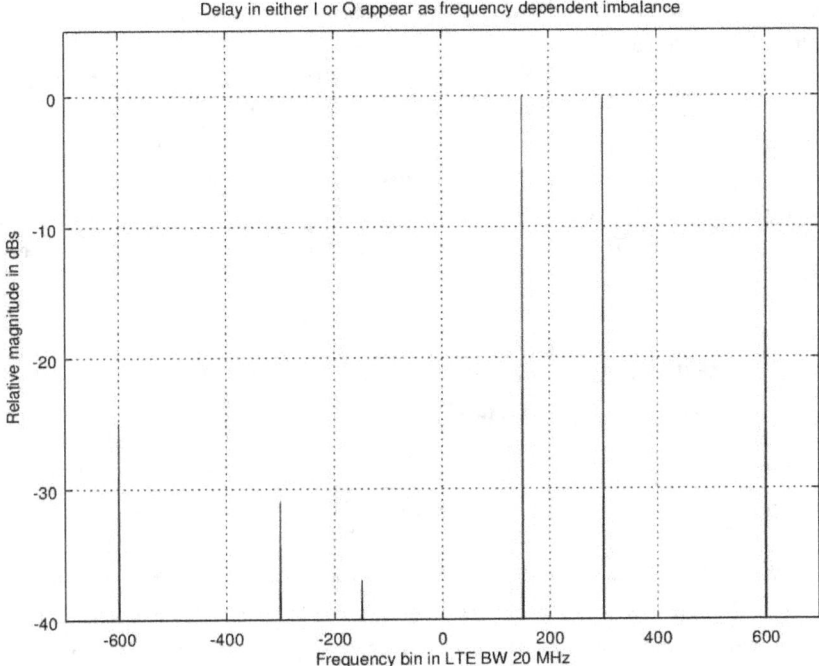

Figure 6.7. Clock skew creating linear phase shift and increasing imbalance (at band-edges) with frequency.

Parameters used in the plot (specific to LTE 20 MHz bandwidth):
Sampling frequency f_s = 30.72 MHz, and Ts = 1/f_s = 32.552 ns
Delay difference (δt) between I and Q signals sampling = 2 ns
Frequency tone or bin spacing = 15,000 Hz
Frequency bin = 150 (frequency ~2.25 MHz), imbalance at bin −150 = −37 dB (−36.993 dB)
Frequency bin = 300 (frequency ~4.50 MHz), imbalance at bin −300 = −31 dB (−30.972 dB)
Frequency bin = 600 (frequency ~9.00 MHz), imbalance at bin −600 = −25 dB (−24.952 dB)
For the completeness of numerical examples, the imbalance in dBs for frequency f in hertz and relative time delay δt in seconds is calculated as 20 $\log_{10}(\pi f \delta t)$ = 20 $\log_{10}(\pi \times 9$ MHz $\times 2$ ns) = 20 $\log_{10}(\pi \times 9 \times 10^6 \times 2 \times 10^{-9})$ = −24.952 dB. As marked in table 6.3, an imbalance of −38 to −39 dB may be a good goal for LTE 64-QAM, and to achieve this, the time delay has to be kept better (lower) than ±0.4 ns, which gives 20 $\log_{10}(\pi \times 9 \times 10^6 \times 0.4 \times 10^{-9})$ = −38.93 dB. As a clarification note, the delay of 0.4 or −0.4 (±0.4) ns introduce the same imbalance magnitude. Practical systems

achieve much lower clock skew than ±0.4 ns to avoid contributions building the noise floor.

Finding the time delay imbalance and also fixing is a difficult task. The delay correction requires time-domain interpolation using Farrow filters or frequency-dependent phase imbalance correction. For LTE baseband signal BW of 9 MHz, the ADC clock difference or equivalent analog signal sampling delay has to be much less than 0.35 ns (that gives close to −40 dB imbalances) by design to avoid getting into measurements and fixing the issues.

6.8   Estimating Rx I-Q Imbalances in Frequency Domain

Considering signal model $x(t) = A\,[(1 + \varepsilon)\cos(\omega t) + j\sin(\omega t + \theta)]$ for I-Q imbalances, in the frequency domain $\dfrac{X(-\omega)}{X(\omega)} = \dfrac{[\varepsilon^2 + 2\varepsilon + j\,2(1+\varepsilon)\,\sin(\theta)]}{[2(1+\varepsilon+\frac{\varepsilon^2}{2}+ \varepsilon\cos(\theta)+\cos(\theta)]} \approx \dfrac{\varepsilon}{2} + j\dfrac{\theta}{2}$ gives directly amplitude and phase imbalance. Wherever possible the preferred approach in Rx imbalance estimation is to go for frequency-domain analysis. By changing the frequency from lower tone index to all the way up to band edge, the phase and amplitude imbalance can be estimated from $\dfrac{\varepsilon}{2} + j\dfrac{\theta}{2}$. The variation in phase imbalance can be used to establish relative clock delay or group delay. In the frequency domain, several tones can be sent simultaneously, such as OFDM signal, but as one side band as shown in figure 6.8. With wide-band signal, the total analysis can be established in one particular operation. The DC offset is propagated to DC or center carrier.

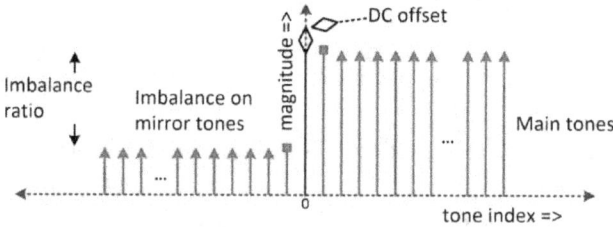

Figure 6.8. Frequency-domain I-Q imbalance estimation from baseband samples.

Even though amplitude and phase imbalances are relatively scale insensitive, care has to be taken in propagating the sign of amplitude and phase imbalances. The DC offset is a magnitude, and in general DC offset accumulates, and sign changes in various cascaded amplifiers, and proper estimation requires closed-loop operation. As an example, keep applying I-Q offsets in a binary search, observe for minimum acceptable value, and finalize the offset used at the analog blocks and digital blocks. In summary, DC offset is not directly used for correction on the basis of the estimated values.

6.9   Estimating Rx I-Q Imbalances in the Time Domain

The key aspect to be used for time-domain amplitude and phase imbalance is the removal of DC offset in both I and Q samples. As explained earlier DC offset can be estimated and cancelled in several easy approaches.

Amplitude imbalance. The amplitude imbalance is estimated depending on I and Q amplitudes over a block of samples. As this is a ratio over block of samples, several approaches can be used, such as power ratio, rms voltage, and mean [abs(N-samples)]. The imbalance obtained can be used either with analog amplifiers or on digital samples. The I-Q samples for amplitude imbalance estimation have to be free from DC offset.

Phase imbalance. The phase imbalance estimation requires samples free from DC offset and amplitude imbalance. In one of the example approach, the formulation can be considered as follows.

$$[AI\,(1 + \varepsilon)\cos(\omega t)]\,[AQ\sin(\omega t + \theta)] = \frac{AI\,AQ}{2}[\sin(\theta) + \sin(2\omega t + \theta)]. \quad (6.6)$$

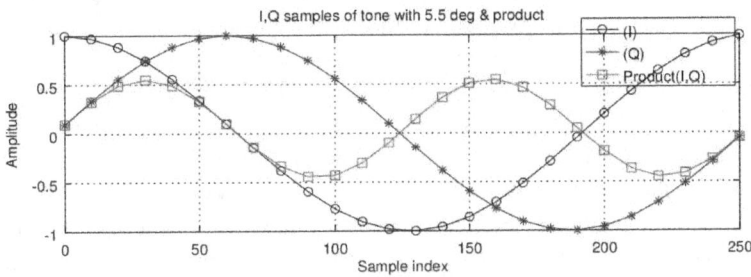

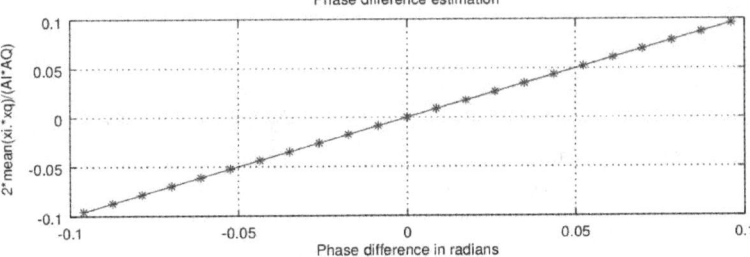

Figure 6.9. Phase difference estimation using sine and cosine.

By integrating (summation) over integer number of ½ cycles, $\sin(2\omega t + \theta)$ goes to zero. For small values of θ, the $\sin(\theta) \approx \theta$. From this formulation, the product of I and Q with I-amplitude AI, and Q-amplitude AQ is given as AI AQ $\theta/2$. Figure 6.9 is shown as 2×mean(xi xq)/(AI AQ), and it gives directly phase difference θ in radians, denoted for I-signal xi, and Q-signal xq. From the above simple formulation, the phase difference estimation is sensitive to DC offset, amplitude imbalance, and requires integration over integer number of ½ cycles. In case the phase term is treated as present in the $\cos(\cdots)$ term instead of $\sin(\cdots)$, the θ estimated has to be reversed in sign as per table 6.1 combinations.

6.10   Imbalance Correction

The phase imbalances obtained are applied in analog and digital processing blocks. As shown in figure 6.10, the DC offset is corrected at the first (usually called coarse) level by subtracting DC in the analog baseband modules. The residue goes through multiple stages of amplification and cancelled in digital processing blocks [URL (limemicro1), URL (LMS7002M)]. The residual DC offset, gain, and phase imbalances are corrected in digital processing as this is difficult in analog and easy to implement on digital samples.

Gain imbalance correction. As explained earlier gain can be applied in either the I or Q paths or in both paths. If gain imbalance is more, the correction will go to both I and Q digital samples. After amplitude imbalance both I and Q will have the same amplitude.

Phase imbalance correction. For an imbalance angle θ in $[(1 + \varepsilon)\cos(\omega t) + j\sin(\omega t + \theta)]$, first the amplitude imbalance is eliminated and after that $I(t) = \cos(\omega t)$ and $Q(t) = \sin(\omega t + \theta)$, and phase imbalance corrected signals are calculated as $I'(t) = I(t) + \tan(\theta/2) Q(t)$ and $Q'(t) = Q(t) + \tan(\theta/2) I(t)$. As $\tan(\theta/2) \approx (\theta/2)$ for small values of θ (where θ is in radians and can take either positive or negative values), the simplified expression is used. Basically a part of $I(t)$ is combined with $Q(t)$, and a part of $Q(t)$ is combined with $I(t)$ creating:

$$I'(t) = I(t) + (\theta/2) Q(t)$$
$$Q'(t) = Q(t) + (\theta/2) I(t).$$

In the diagram in figure 6.10, amplitude imbalance correction is shown first and then phase imbalance correction. For small values of imbalances, the order of these two corrections may be switched.

RF → Baseband amp/filters → I-Q samples → DC offset comp → [↓2] → Gain comp → [↓2] → Phase comp → I-Q samples

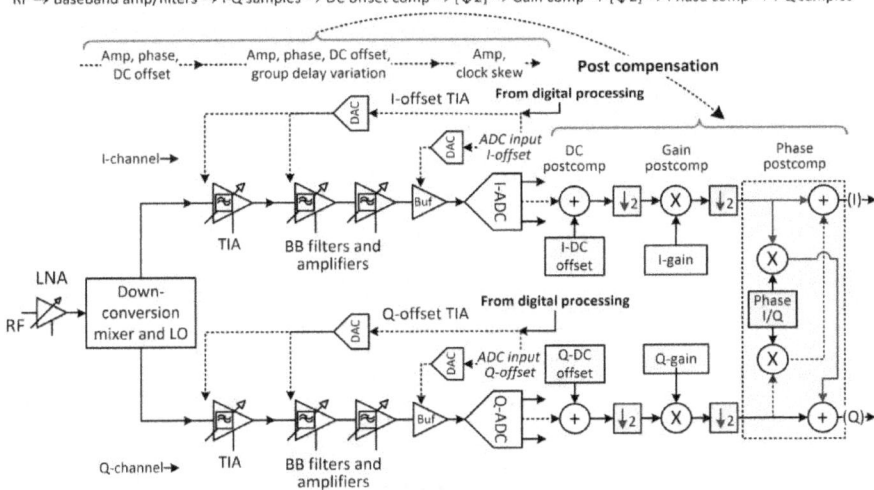

Figure 6.10. I-Q imbalance corrections in analog and digital operations.

6.11 LTE BW and Band Edges

Table 1.1 lists LTE BW, RBs, tones, and interdependent basic time-frequency parameters. The control signal tones are present at the band edges. For example, the physical uplink control channel (PUCCH) that carries uplink control information (UCI) uses the tones on either side of the band edge. During regular symbols the band-edge tones will be with PUSCH data, and in that situation also, it is essential to preserve the band edges. In general, it is essential to preserve the frequency response flatness up to the band edges, and the specifications of baseband analog and digital filters have to be considered with care to create only small loss of the order of ~1 dB or less for the band-edge tones. This is applicable to all the LTE BWs of 1.4, 3, 5, 10, 15, and 20 MHz. The band edges refer to the occupied BW; as an example for LTE 10 MHz, occupied RF BW is 9 MHz, and baseband signal band edge is at 4.5 MHz.

7 TRANSMITTER I-Q IMBALANCES, ESTIMATION AND CORRECTION

In the previous chapter, receiver (Rx) in-phase and quadrature-phase (IQ or I-Q) imbalances/impairments referring to baseband signal are presented, and in this chapter, I-Q imbalances are presented for the transmit (Tx) path. Equation (7.1) includes four types of I-Q impairments, the amplitude error ε, phase θ, DC offset in I (Ioff) and Q (Qoff), and time delay or clock skew between I and Q as δt. The Tx imbalance estimation and cancellation in this chapter are focused on amplitude, phase, and DC-offsets. As described in Rx impairments, the Tx impairments are also majorly contributed from analog and radio-frequency (RF) sections of hardware:

$$x(t) = A\left[\{\text{Ioff} + (1 + \varepsilon)\cos(\omega t)\} + j\{\text{Qoff} + \sin(\omega(t + \delta t) + \theta)\}\right]. \quad (7.1)$$

As a starting point of Tx I-Q imbalance representation, the samples going to the digital-to-analog converter (DAC) are imbalance free as these samples are generated in digital signal processing blocks. In case the DAC is of current source, the hardware uses cascaded transimpedance amplifier (TIA) to convert current to voltage and also to act as first-level image rejection filter. For generic representation and continuity of the description, baseband TIA is included in the subsequent sections. The I-Q DACs stage introduces two main components of imbalance—clock skew that is a linear phase with frequency, and amplitude imbalance between two DACs, followed by gain (amplitude) and phase imbalance in TIAs. The imbalanced analog signal (from DAC or DAC+TIA) goes through multiple stages of amplification and filtering, and fed to up-conversion mixer stages. As marked in figure 7.1, the dominant imbalances are amplitude, phase, DC offset, and group delay variation. The end result of imbalances appear on the RF signal that goes through further amplification in cascaded amplifiers—referred to as power amplification (PA). As an industry established practice, the I-Q imbalances are not corrected in analog and RF blocks—except in commercial RFICs [URL (AD9361), URL (Max2837)].

The digital processing provides more flexibility for LTE multiband systems automation and this is the industry established common practice for imbalance correction. Essentially, to fix imbalances, the imbalances are estimated or measured and propagated back to time-domain digital signal processing for precompensation (used short form precomp in this book).

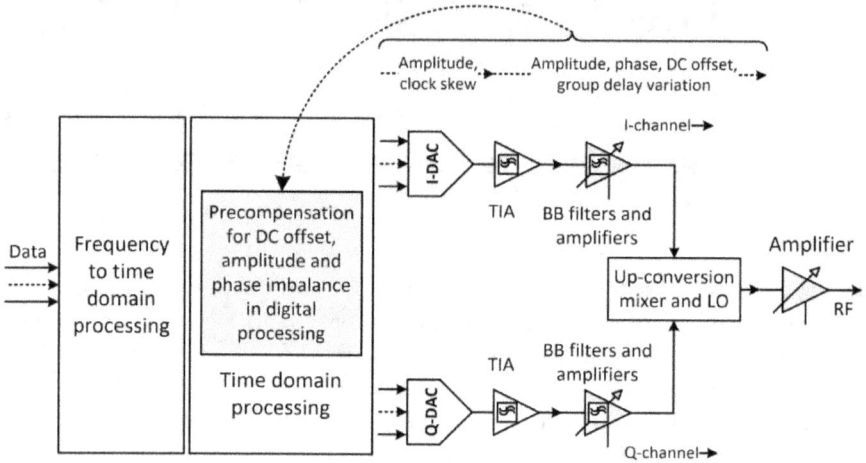

Figure 7.1. I-Q imbalances in Tx.

Once the imbalances are known, the precomp on time-domain samples is relatively easy with the caution of side effect of reduction to the DAC dynamic range by a few dBs. The phase imbalance that requires treating I-Q signals together as explained in next sections. DC offsets are independent in I-and-Q. The scaled DC offsets are added in the digital.

The amplitude imbalance adjusts the gain of I relative to Q, or both I-Q signals can be adjusted with geometric mean. As an example, if I channel is 0.8 times of Q in analog signal path, the I_{dig} is either scaled up by $1/0.8$ ($=1.25$) or Q_{dig} is scaled down as 0.8, or both I_{dig} and Q_{dig} are scaled with one common geometric mean sqrt(1×0.8) = 0.894. Using geometric mean, the I-path I_{dig} is scaled up as $I_{dig}/0.894 = 1.118\ I_{dig}$ and Q_{dig} is scaled down as $Q_{dig} \times 0.894 = 0.894\ Q_{dig}$. To represent this example in dB scale, the gain imbalance is $20\ \log_{10}(0.8/1) = -1.94$, and for this, half of 1.94 dB is distributed to I and Q paths. The I-digital path gain is increased (signal scaled up) by 0.97 dB, and Q gain is decreased (signal scaled down) by 0.97 dB. The cascaded $1.118\ I_{dig}$ with analog imbalance of 0.8 gives $0.894\ I_{Ana}$ that matches with Q_{Ana} that is already scaled down to 0.894 on digital samples.

7.1   Amplitude Precomp Reduces the DAC Dynamic Range

The original input without imbalances may be properly scaled to match DAC input range of 12 bits number format that is equivalent to 1 V pk-pk at DAC output. Taking example of making I-amplitude to correct for 0.8 in analog path, the digital samples amplitude is scaled up by a factor of 1.25. The full scale at DAC output as per math is 1.25 V pk-pk. If DC offset to be added is 0.1 V (to nullify equivalent of −0.1 V in analog), the positive peak moves to $1.25/2 + 0.1 = 0.725$ V. The allowed DAC normalized value is 0.5 V for 1V pk-pk. The digital samples in I and Q have to be backed off (reduced in amplitude) before applying I-Q amplitude and DC offset imbalance. The dynamic range loss in the example is $20 \log_{10}(0.725/0.5) = 3.23$ dB. To account for the growth of signal, the original imbalance free samples in I-Q have to be intentionally reduced (scaled down) in amplitude by ~3.23 dB.

By distributing amplitude in I-Q with geometric mean, instead of scaling I path by 1.25, the scaling is 1.118, and with offset the positive peak is $(1.118/2) + 0.1 = 0.658$ and the loss is $20 \log_{10}(0.658/0.5) = 2.4$ dB, lower from 3.23 dB of I-only path correction. From this interpretation, for large amplitude imbalances, the favorable approach would be to distribute the amplitude imbalance in both I and Q. Summary here, create a headroom in the time-domain samples that are going to DAC I-Q preimbalance correction. The headroom can be adaptive or created for the worst-case imbalance values. For higher amplitude imbalances, distribute the imbalance in both I and Q, and such distribution modifies the gain in one path (I-path in the above examples) and loss/attenuation in another (Q) path.

7.2   I-Q Imbalance Correction by Design

I-Q imbalance estimation and correction evolved in multiple ways. If the system is stand-alone and of high value instrument or simulator, the lab characterization brings out several parameters to build the gain and frequency-dependent tables that can be used in offline to keep using the hardware/instrument at the required performance levels.

Several RF transceivers for UE and eNodeB are made with better balance that will not call for imbalance measurements for the typical required system performance. These are of commercial nature RF transceivers [URL (AD9361), URL (MB8613L13A), URL (Max2837)], and the manufacture tunes the chips by design, during fabrication and packaging, and preloads or fuzes the right device-dependent parameter tables that will give required default performance. The extra device options will help improve the performance through programming. Refer to I-Q imbalances in [URL (Max2837)] that listed group delay and group delay ripple. For I-Q time delay imbalance what matters is the group delay difference between I-Q channels that translates to frequency-dependent (often referred to as frequency-selective) phase imbalance.

With a small form-factor (size) and highly integrated devices, several optimization factors can be attempted in terms of optimization for silicon area, design time and cost, power dissipation, and time to market. Such RF transceivers are not fully tuned for optimum imbalance reduction by the design and manufacturing process. The optimization for performance can be achieved by self-calibration cycles in a final product, which may precede the actual usage of the device for higher electrical performance.

7.3   Tx Imbalance Digital Correction Architecture

As marked in figure 7.2, the I-Q imbalances start from the DAC stage and ends at the mixed output in RF. The imbalances are precompensated in digital before feeding samples to the DAC.

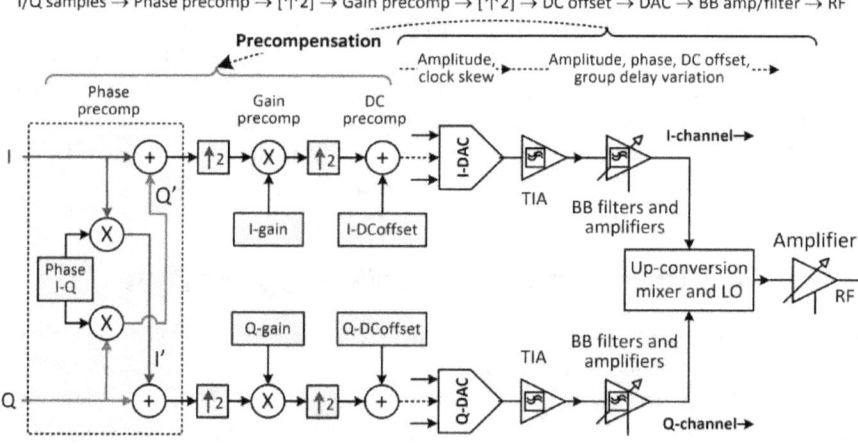

Figure 7.2. I-Q imbalance precompensation in digital processing.

Taking an example of LTE 20 MHz BW operation, the samples from frequency to time domain are sampled at 30.72 MHz. These input samples are up-sampled to benefit the DAC output baseband filtering. The industry established practice of up-sampling is by a factor of 4. The digital processing chain also includes a generic gain correction to adjust for creating the headroom for peak-to-average-power ratio (PAPR) margin and to create headroom for DAC input signal growth due to imbalance precompensation. The typical precompensation is shown in figure 7.2 [URL (limemicro1), URL (LMS7002M)]. In the digital processing, the time-domain samples from frequency to time domain are subjected to first phase imbalance *precomp*. As phase shifts are usually low, the signal amplitude will not grow significantly. Next, amplitude imbalance is compensated with fine gain changes in I or Q or I-Q paths. The amplitude imbalance correction can be combined with the usually digital gain present in the digital processing chain. The DC offsets are independently compensated in I-Q, and this may be the last operation of precompensation. The imbalance correction operations are cascade with the interpolation stages. The imbalances corrections and interpolations can be either distributed as shown in figure 7.2 or corrected first before interpolation, by following the sequence of imbalance operations as phase precomp → amplitude precomp → DC offset precomp.

7.4   Time Delay Correction

As explained in the previous chapter, a relative time delay or clock skew may happen at DAC stage. To summarize, if perfect samples of functions $\cos(\omega t)$ is fed to I-DAC and $\sin(\omega t)$ to Q-DAC, the output of DACs appear as $\cos(\omega t)$ and $\sin(\omega(t+\delta t))$ [both analog and samples are left in analog signal equivalent representation for easy mapping] a delayed version of sine wave in Q (with reference to I-path) that creates frequency-dependent phase imbalance. The group delay variation in analog filters would not be dominant for most application as these filters are of higher BW than useful signal BW. If such group delay variation exists, the industry followed most common approach is to arrive at a fixed value that will be optimum across all the frequency components. With reference to figure 7.2, the Farrow filter can be cascaded or combined with the interpolators that can preadjust the delay in I or Q or in both I and Q. As these Farrow filters [URL (Farrow1), Edmund Coersmeier] are more complex to implement, effort will be made to eliminate such time delay impairments to acceptable levels in the design of DAC clock buffers or clock routing, and analog filters group delay match between I-Q channels. Due to four times over sampling ratio (OSR) at DAC, the analog filters can be designed with minimal relative time delay variation between I-Q channels.

7.5   Manual Characterization of I-Q Imbalances

In the manual correction of imbalances, the RF output is connected to a spectrum analyzer or a LTE friendly PXA analyzer [URL (PXA)]. The imbalances are controlled through a computer interface of user equipment (UE). The imbalance control settings/registers can be modified through computer-controlled interface that could have graphic user interface (GUI). As explained in previous chapters, the DC offset appears as a center carrier at local oscillator (LO) frequency. The I-offset is manually adjusted by increasing from zero toward positive value of the offset. If the Tx LO feedthrough (*LOFT*) is decreasing, the steps will continue until seeing local minima. If Tx LOFT is increasing the operation will be continued in opposite direction. Next, the similar steps are repeated for Q offset for achieving some more reduction in Tx LOFT.

The amplitude imbalance adjustment requires at least 1 tone of I-Q digital samples going through up-sampling, and DAC. The preferred way is to use 3 to 5 points spanning the signal bandwidth (BW). One-sided OFDM signal can also be generated that can give complete half BW imbalance. For adjusting, the imbalance can be corrected in I or Q or both depending on the imbalance level. If the imbalance is below 10%, correction in only one path may be used.

Next follows the phase imbalance correction. Phase imbalance goes to both I and Q together. The whole process is performed at a particular analog signal path gain. Next move on to valid analog signal path gain and repeat the steps. In Tx, only a few steps of analog gain is used, usually to help improve the nonlinearity adjustments of up-conversion mixer. At the end of experimentation, a table has to be built for the analog baseband section gain steps.

Next comes voltage and temperature corners, and wherever the analog signal path imbalances differ and it is essential to get more calibration values. The manual approach explained here is entirely with human interaction. This approach is aiming at obtaining minimum value with manual sweep. This is good for initial functional characterization of device and to obtain limits of imbalance, and to make sure imbalance corrections are working and stable in an open loop—to achieve required specifications.

7.6   I-Q Imbalances Estimation with BiST

In the manual correction of imbalances, the RF output is connected to a spectrum analyzer or a LTE instrument. In closed-loop operation, the RF output from Tx is looped back to another Rx (in place of an instrument) with additional hardware and software options.

The technique built in self-test (BiST) is given in [URL (BiST), Afsaneh Nassery et al. (2012)] that sends multiple combination of 2 tones as $\cos(\omega_1 t)$ and $\sin(\omega_2 t)$ and derive parameters through analysis. Note that the tone in I-path baseband is at f1 and Q-path is at a different frequency f2. The Tx will have 2 tones at f1, and f2 offset from the Tx LO frequency. The detector is assumed to be linear or operating in the linear region. The baseband receiver analyses the signal and extracts signal strength at DC, f1, f2, 2f1, 2f2, f1−f2, f1+f2 that are used in building equations on the basis of the signal strength that allow mathematical computation of the I-Q imbalances.

7.7   Tx LOFT and DC Offsets

Tx LOFT is the leakage of Tx LO through various paths. In normal operation I-Q baseband signals are up-converted to the required frequency decided by the LO frequency. The LO leakage is undesired component that should not be transmitted on air interface. Once LOFT is created it becomes difficult to remove through RF filtering. LTE 3GPP standards [3GPP 36.521-1, 3GPP 36.101] defined the limits for acceptable LOFT levels. In LTE, resources are shared by multiple UEs, and the Tx LOFT of UE transmitters interferes with other UE transmissions.

The LOFT has to be controlled by design in RF sections. With multiple bands operation, meeting the specifications with RF alone perfections will be limited. The industry established approach is to include finer DC offset control along with I-Q digital samples. The DC offset with varying amplitude and polarity cancels (or technically precompensates the LOFT). To cancel the LOFT, the calibration operations estimate the LOFT and, in a closed loop, establish the required DC offset, and store several tables in factory settings or incrementally improve run-time estimates.

7.8   Tx LOFT Estimation with Tx to Rx Loopback

In the BiST-based approach, separate receiver is used that takes the Tx detector output and analyzes for frequencies and signal strength. In an alternate approach, a part of RF Tx output is fed back to one of the LTE receiver and analyzed for the imbalances.

Considering Tx LOFT, it appears at Tx LO frequency that has to be cancelled by precompensating the Tx path DC offsets on I-Q digital samples. In LTE-FDD, the Rx LO frequency is away from the Tx LO due to frequency duplexing, and for this reason the Tx LOFT fed to the Rx will get attenuated through the receiver. In LTE-TDD, Rx LO is the same as Tx LO. If TDD-Rx is used in loopback to measure Tx LOFT, the Tx LOFT gets converted to DC and gets combined with the Rx DC offsets.

To extract Tx LOFT using loopback mechanisms, the Rx LO has to be offset by known frequency δf. With this approach, the Tx LOFT goes through Rx and will become a component at δf. The signal strength in Rx at δf has to be used in estimating the Tx LOFT and finally required Tx chain DC offsets have to be incorporated to cancel Tx LOFT.

The equivalent signal strength at δf in Rx is not directly related to the precompensation value in Tx digital processing. A simplified approach is to go for a (binary) search by starting with the Tx chain precompensation DC offset in the expected bounds in I, and the I-channel DC offset is varied and Rx signal strength at δf is minimized. Next DC offset in Tx precomp Q-channel is varied and signal strength minimum point is recorded to get finer correction.

In the estimation of signal strength at δf, the DC offset from Rx chain, and image component of δf due to Rx amplitude and phase imbalances have to be isolated. In the frequency-domain-based analysis, the DC, δf, and image component of δf are distinct.

In time-domain-based signal strength estimation, the DC, and image of δf have to be removed before estimating the signal strength. In general the signal strength from image of δf is usually ~20–24 dB lower than the main component at δf, which is insignificant for this closed-loop measurements. Thus the removal of Rx-DC component is essential that leaves the main component at δf and its insignificant image.

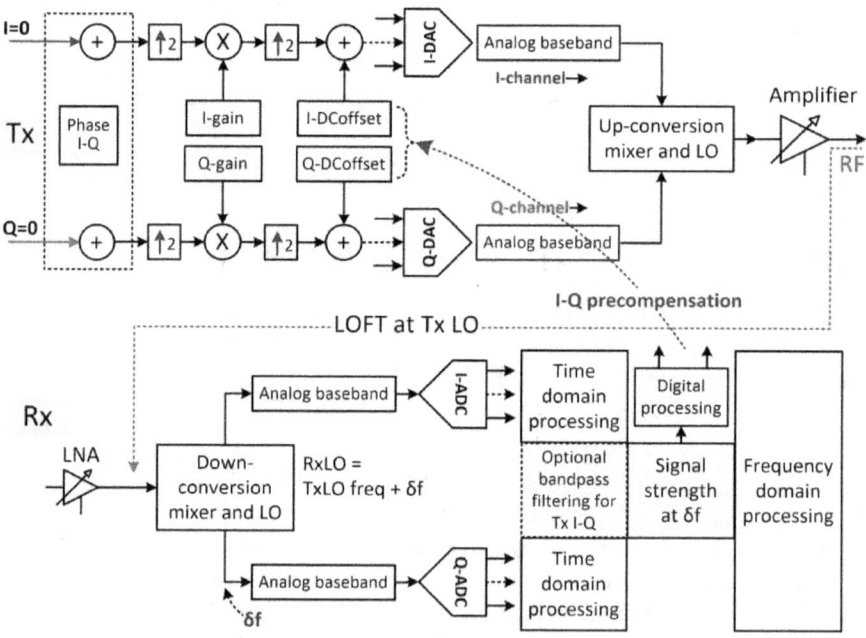

Figure 7.3. Tx LOFT estimation through Tx to Rx loopback measurements.

As a summary, even if Tx I-Q samples are zero, the Tx LOFT is present. By looping Tx LOFT to Rx and looking at signal strength in Rx, the required DC precompensation at Tx can be estimated. While performing this test, Rx uses a set of filters with BW, and amplification settings. As such, it is not possible to directly conclude on how much of Tx LOFT is present or did it reach required levels of LOFT rejection. An offline characterization of Rx is essential that provides Rx signal strength at δf verses Tx LOFT levels. If Rx is used for 3–4 gain and filter settings, such mapping of Rx signal strength at δf verses Tx LOFT has to be established for multiple settings.

7.9   Tx Amplitude and Gain Imbalance with Tx to Rx Loopback

Using Tx to Rx loopback mechanisms, Tx gain and amplitude imbalances are implemented by sending a digital samples tone at frequency fb from I/Q digital baseband, and Tx RF will be created at TxLO+fb as main strong component, and TxLO-fb as imbalanced image frequency component, and Tx leakage or LOFT at TxLO frequency (figure 7.4).

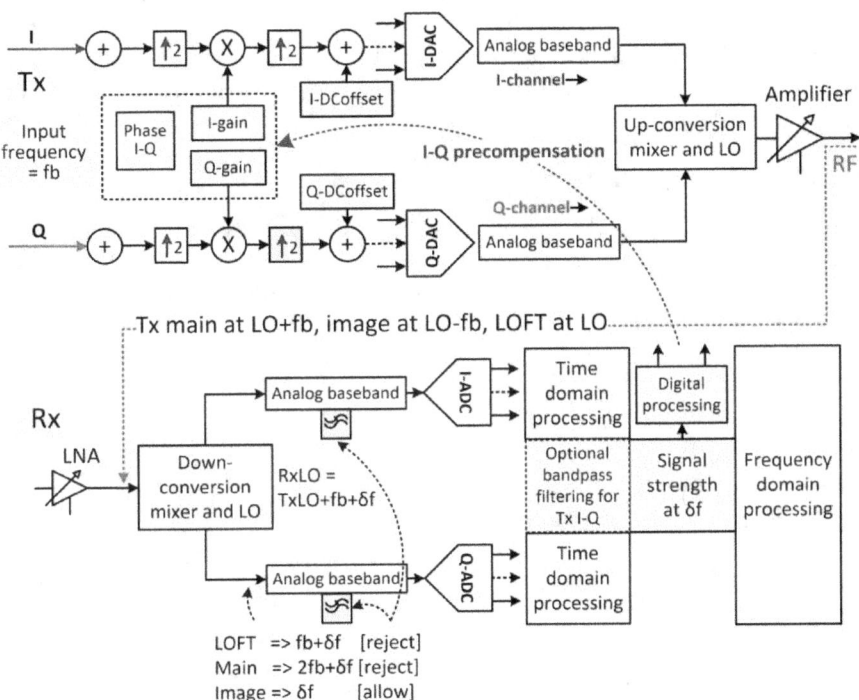

Figure 7.4. Tx gain and phase imbalance correction through Tx to Rx loopback measurements.

The Tx RF signal when looped back to Rx, it will get down-converted and passed on to the Rx-baseband analog and then to digital samples, and finally to time-domain and frequency-domain processing. Rx can include its own imbalance; hence, it is essential to reject main component and LOFT component and pass on only image part. The baseband with Rx filters will preserve δf and passes on to the baseband DAC. The Rx can also create its own imbalance at $-\delta f$, which is insignificant. The requirement is to measure the signal strength δf.

The I-Q imbalance estimation starts with zero precompensation in amplitude and phase. The amplitude is set to multiple values in a binary search and the minimum signal strength at δf is established. Similarly, the phase imbalance is also varied to arrive at the minimum signal strength at δf. The established gain and phase imbalance will be entered into a table that corresponds to the selected Tx setting. For a different Tx gain/attenuation and filter settings, the above steps have to be repeated.

When frequency-domain processing is used, the requirements on baseband filtering will be less demanding. The Tx may send multiple tones instead of 1 tone at fb, and frequency-domain analysis can simultaneously extract useful information including for group delay variation. Note that for frequency-domain analysis by reusing LTE-FFT blocks, the tone spacing has to be multiples of 15 kHz; that way the frequency components can maintain orthogonality.

7.9.1   Tx Group Delay

As an extension to the phase imbalances, clock skew or group delay imbalance can also be derived at coarse level. For this, input fb has to vary to multiple frequencies, and if this is returning multiple phase imbalances, an estimate can be made for group delay.

8 TRANSMITTER GAIN MANAGEMENT AND TIME ALIGNMENT

In this chapter three topics are presented: (1) transmit (Tx) dynamic range, (2) gain control/management, and (3) timing alignment specific to LTE UE (user equipment) Tx path that relates to uplink (UL) signal quality. The Tx gain control modules have to be tied for timing alignment to improve signal quality. The LTE system operates with a time resolution interval of Ts, which is 32.55 nanoseconds (ns). In the case of sampling clock offset (SCO) of chapter 5, the SCO timing errors will be controlled to a small fractional of Ts. Referring to chapters 6 and 7 on in-phase (I) and quadrature-phase (Q) signal imbalances, the timing delay imbalance in I-Q will be in nanosecond scale, and will be much lower than Ts interval. For UL gain control, the timing alignment at the boundary of gain/loss change is important and has to be maintained to microsecond scale across multiple modules of sample processing, mixed signal, analog, RF, and RF FEM (front end module). The alignment mismatch has to be lower than cyclic prefix (CP) of the symbol for improving quality at symbol and subframe (SF) boundaries.

8.1 Power Control with SRS in FDD and TDD

Considering example representation in figure 8.1 for FDD subframes with sounding reference signal (SRS), the following observations can be made:
- The SF has 14 symbols (chapter 1).
- SRS when scheduled occupies the last symbol of the subframe.
- The SRS can be of different power than the other 13 symbols of the subframe.
- In general, LTE UE has to possess capability to accept power change—increase or decrease from the previous SF—at SF boundary.
- By imposing SRS power control requirements, SRS power can be

higher than its own SF power, subjected to staying within +23 dBm maximum limit.

- SRS power can be lower, subjected to staying ≥−40 dBm.
- At the end of SRS (end of full SF), the next subframe can be of a different power level.
- SRS can be blank, and during blank period, the power has to be −50 dBm, and next SF can start at the end of the blank.

Notes on SRS: SRS is sounding reference signal tones present in the SRS symbol. SRS appears on 4–96 RBs with different frequency offsets and frequency hopping patterns. The parameters of SRS are configured from eNodeB, and a few parameters are internal to the UE. LTE has two modes FDD and TDD. In FDD SRS is on symbol 13 (last symbol) of a selected subframe. In TDD it can appear in the last symbol or in other symbols of a special subframe (SSF) depending on various TDD configurations (chapter 1). SRS also uses the same power limits of −40 to +23 dBm during active transmission, and less than −50 dBm during SRS blanking. SRS symbol power levels can be much higher or much lower depending on the various parameters exchanged between eNB and UE and SRS power control. In practical tests with eNodeB, SRS power difference of −6 to +15 dB is observed with reference to physical UL shared channel (PUSCH) power. The SRS power difference with reference to PUSCH can make RF FEM PA (power amplifier) to change the gain by big steps, which requires special effort to time align for improving LTE signal quality. *Note, the illustrations in figures 8.1 and 8.2 are for functional representation to covey the scenarios.*

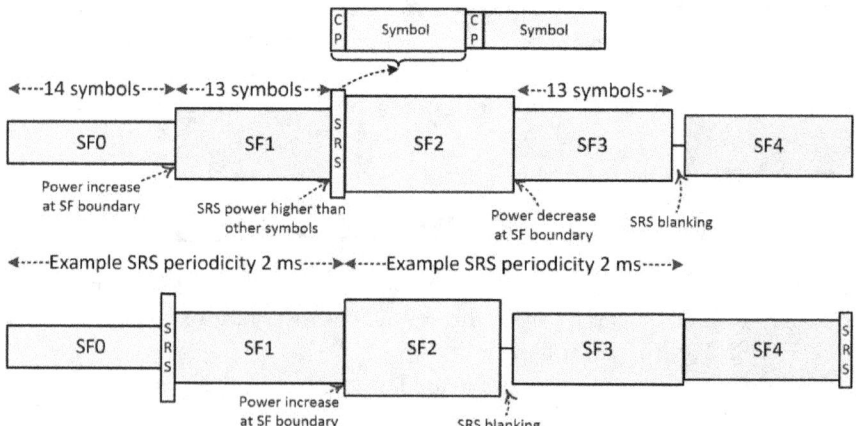

Figure 8.1. LTE FDD UL subframes with SRS, and power control. In figure, x axis is time and y axis represents power level.

Considering the example shown in figure 8.2 for TDD subframes with SRS, we can observe the following:

- The TDD can schedule SRS in special subframes and normal subframes.
- Refer to SF1 in figure 8.2 for SRS in special subframe; the SRS can be present in the last 2 symbols in any combination of presence and blanking—shown in figure 8.2 for 3-different scenarios.
- In nonspecial or normal subframes, the high-level SRS behavior of presence with multiple power levels, and blanking are similar to FDD examples.

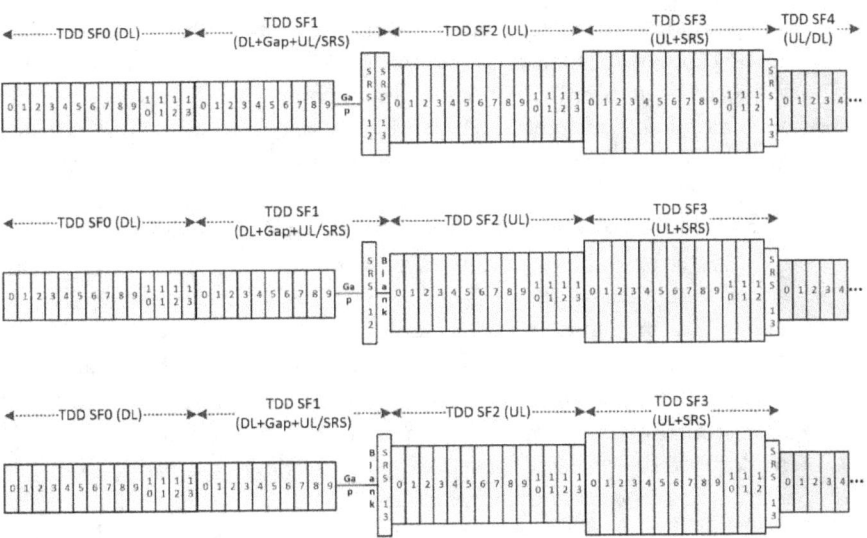

Figure 8.2. LTE TDD UL subframes with SRS, and power control.

8.2   Tx Chain Gain and Loss Distribution

Referring to figure 8.3, under normal operation—even with SRS presence—the gain/loss distributions summary is listed below:

- The average signal level at IFFT output is steady in some normalized reference. The digital I-Q samples are created with headroom for PAPR as described in chapter 3, and I-Q precompensation (chapter 7).
- The sample processing uses coarse gain mainly for the following:
  - Under rare situations when SRS is using much higher power difference (say, 12–18 dB or more), then a few dBs of digital coarse attenuation/gain may be used depending on the capabilities and power change transient response of RF and RF PA gain/attenuation control.

- When network signaling is present to improve emissions, 1–3 dB gain reduction may also be used as digital attenuation control without altering the analog/RF gain.
- Power back-off on baseband digital samples may help to control the nonlinearity of the RF mixer/up-conversion stages. This is to meet certain spectral mask in selected bands, and to reduce amplitude during SRS blanking.
- The sample processing incorporates fine gain control for the following:
  - Residual gain correction, if gain steps in analog/RF are not in acceptable limits.
  - I-Q gain imbalance precompensation (see chapter 7).
- The I-Q digital to analog converters (DACs) are utilized to full scale after accounting for PAPR and headroom margin.
- The analog baseband keeps gain and filter stages. Similar operations of digital sample processing coarse gain control can be used in analog, but will be restricted to a few selected steps.
- The RF gain control happens at RF pre-PA and PA (chapter 2) stages:
  - LTE system requires a minimum power control from −40 to 23 dBm a total range of 63 dB in 1-dB steps. The 1-dB step is required to incorporate transmit power control.
  - When Tx has to be blank with gain control it should have capability to provide below −50 dBm. Thus making the requirements to 73 dB gain control. In many implementations, the blanking is partly implemented as digital attenuation of samples instead of entirely using RF, and RF PA dynamic range.
- There will always be a few extra dBs margin to redistribute gain/losses in RF FEM while handling multiple bands, and multiple PAs.
- The PAs provide typical gain control of 24–28 dB in finite steps, and PA driver (which is part of RFIC) provide another 40–50 dB control (usually attenuation 1-dB steps), and digital samples are controlled by another 0–12 dB in coarse and fine gain steps.

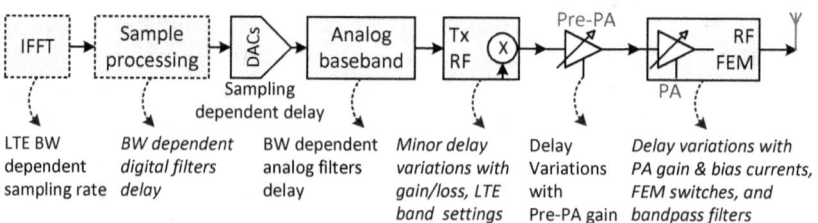

Figure 8.3. LTE Tx flow with delay variation markings.

8.3   Timing Delay Variations in Tx

As presented in chapter 2, and marked in figure 8.3, the Tx time samples and timing relations are established after inverse fast Fourier transform (IFFT) block. As described in chapter 1, the sampling at IFFT output varies as per the LTE channel BW or number of IFFT points and both are related. As explained in chapters 6 and 7, these samples will go through up-sampling by a factor of 2–4 (usually 4) and also incorporated with I-Q imbalance precompensation. The up-sampling helps relax the frequency response of analog filters (part of analog baseband) present after DAC. The analog filters and gain stages condition the I-Q DAC outputs, and I-Q signals go to radio-frequency (RF) up-conversion stage. The RF signal goes through pre-PA (preamplifier to PA also called PA driver) and finally to PA stage. Refer to chapter 2 for some more details on the signal flow.

As shown in figure 8.4, considering time-domain samples as reference, the SF boundary is at time mark t_0. The sample processing has several stages of filtering, and interpolation that moves the SF boundary to time mark t_1. The analog baseband introduces group delay and moves the SF boundary to time mark t_2. The pre-PA and PA output are marked with t_3, and t_4. Under steady gain—meaning no gain change at SRS symbols or at SF boundary starting from digital samples to PA, then there are no concerns on the delays and time marks from t_0 to t_4.

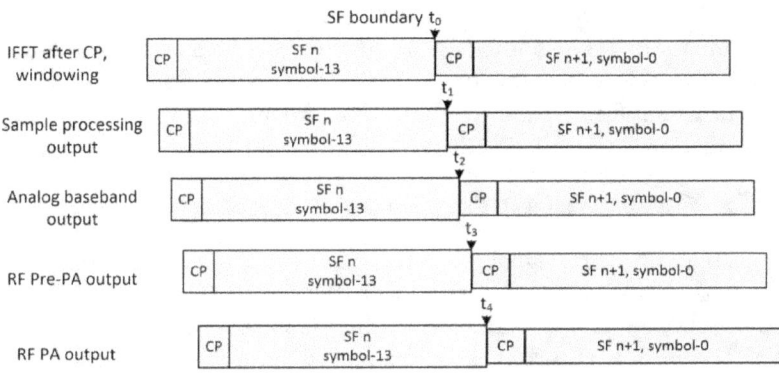

Figure 8.4. Propagation of SF alignment across multiple stages.

Notes: When delay changes in the Tx path, the eNodeB sends timing advance commands, and UE applies and corrects the timing; this works in close-loop operation with eNodeB. Then the question arises why to discuss on this subject—timing alignment across multiple modules of Tx chain?

The timing alignment relations are required to rightly time
- when to apply gain control to digital samples, analog, RF;
- when to apply precompensation for I-Q imbalances, DC offset

> precompensation for local oscillator feedthrough (LOFT); and
> • when to control RF front-end module (RF FEM) and switches.
> LTE transmissions are configured on SF basis (1 ms), and some parameters
> are applied to hardware on per symbol basis—in case the SF contains SRS.
> The transmission quality can be improved by controlling timing
> relations/alignment to be lower than CP time interval.

When gain or losses are changing at SF n to SF $n+1$ boundary, then applying the changes at t_0 time mark for all the stages will distort the SF n tail-end of symbol 13, which is not allowed as per 3GPP. The gain ramp-up delay/response time is allowed only at the beginning of the SF $n+1$ in the region of symbol-0 CP. With reference to the above description, for optimal usage the time align of gain/losses across multiple stages have to apply parameters to take effect at t_0, t_1, ..., t_4 in figure 8.4 examples. How these time marks are arrived? The parameters will come from design, simulations and characterization of the hardware and firmware under various valid use case conditions.

As marked in figure 8.3, these parameters will vary on the basis of LTE BW selection—such as in sample processing, mixed-signal stage and analog baseband. The parameters will also vary on the basis of gain in pre-PA and PA. As marked in figures 8.1 and 8.2, the time alignment gets more difficult when SRS, and PUSCH are operating at different power levels and power is changing at SF boundary. Section 8.1 has some examples on such power change combinations. The summary notes from figures 8.3 and 8.4, and this section example is to apply the time aligned parameters to minimize the distortions at the subframe and SRS symbol boundaries.

8.4   Tx Time Alignment Programming

Referring to figure 8.3, the signal path starting at IFFT goes out as real time operation. Various architectural break-up of BBIC, RFIC, and RF FEM are presented in chapter 2. A condensed replica from chapter 2 is given as figure 8.5. In the Tx control or programming path, the parameters go through various interface options as listed below:
- BBIC programs RFIC through SPI or DigRFv4.
- BBIC or RFIC programs the RF FEM through MIPI-RFFE, and GPIOs.
- The real-time signal going through various stages of figure 8.3 has to match the timing for when to apply gain to digital, analog, and RF modules, precompensation for I-Q imbalances, DC offset precompensation for local oscillator feedthrough (LOFT), and RF switch controls.

- As per 3GPP 36.521-1 and 36.101, the ramp-up at subframe boundary and SRS symbol boundary can utilize ~20 µs, and depending on various other symbol conditions, ramp-up goes up to ~40 µs. In practice allowing such high ramp-up time distorts the data in data symbols or SRS symbol. This is where the signal quality comes into picture to tightly control the ramp-up to less than symbol CP duration of 4.7–5.2 µs.
- With the advancements in the technology, many modules used in LTE system are controlled to µs scale. The RFIC and PAs are able to maintain 1–2 µs of ramp-up. A fine-tuned system may provide a ramp-up of 3–4 µs, which is less than CP duration.

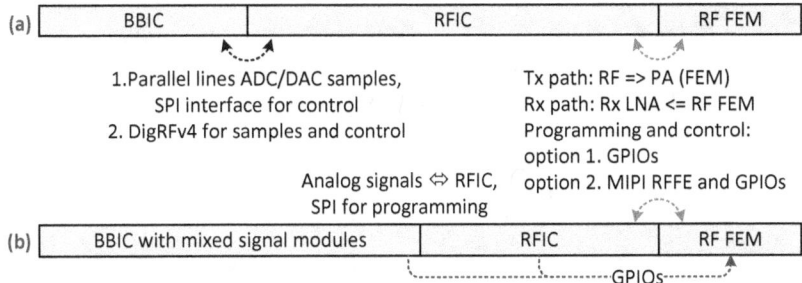

Figure 8.5. Architectural split of digital baseband mixed signal, analog baseband, RF, and RF FEM showing signal and programming path. (a) Conventional architecture with BBIC, RFIC, and RFIC FEM split. (b) Mixed signal and parts of analog inside the BBIC.

8.5   Chapter Summary

The focus in this chapter was on the dynamic range, how the distribution of gain control can happen, and how to time align various signal levels so that the ramp-up/down time is much lower than the CP, which is possible with proper alignment of multiple stages programming/control plane in the signal path. As delays in the signal path across multiple stages are different, care is required to preprogram certain parameters to act at a particular time stamp. For example, DigRFv4 has multiple time alignment strobes (TAS) that are programmed, and functionally the parameters are applied after reaching the TAS value.

9 RECEIVER AGC AND BLOCKERS

In the receiver (Rx) path, the downlink (DL) radio-frequency (RF) signal from the antenna and the RF front-end module (RF FEM) is amplified in the low-noise amplifier (LNA) and down-converted to the baseband signal. The analog baseband signals as in-phase and quadrature-phase (I-Q) paths are further enhanced in the analog baseband and sampled in analog-to-digital converters (ADCs). The digital I-Q digital samples are processed in the LTE baseband and higher layers. The baseband is associated with a processor that controls and programs the RF, RF FEM, and other time-dependent activities.

9.1   Rx Automatic Gain Control

Receiver automatic gain control (AGC) steps and salient points are listed as follows:

- The received signal enters the antenna port and goes to LNA through RF FEM. The FEM (chapter 2) is usually of the fixed loss stage between the antenna port and LNA. The loss of FEM slightly varies depending on the band of operation.
- The desired Rx signal, undesired blocker signals, and transmitter (Tx) leakages—in the case of frequency-division duplex (FDD)—enter the Rx LNA. Both Tx leakage and blocker signals are distinct from the desired LTE Rx signal but adjacent in frequency from the Rx signal.
- The FEM bandpass filters or duplexers attenuate the blockers to some extent depending on the band, and even after filtering, a significant undesired energy enters the LNA.
- Tx leakage is mainly decided by the duplexer isolation between Tx and Rx.
- LNA matching can be treated as wide-band bandpass filter and attenuates the Tx leakage and blockers by a few dBs.
- LNA and down-conversion mixer stages see the main signals as desired Rx, undesired Tx leakage, and blockers. The linearity of these

stages is important for maintaining the Rx signal quality.

- Often mixer output will have a low pass filter with an amplifier that attenuates partly Tx leakage and blockers.
- The analog baseband operation consists of variable gain and variable filter bandwidth (BW) to adapt optimally to the LTE-selected channel BW. The filters will preserve the Rx signal and partially attenuate the undesired signals.
- In the LTE system, as ADCs are of over sampling with typical over sampling ratio of 4, the analog baseband filters are of more wider BW than LTE-occupied BW. Hence, a part of Tx leakage can cross the ADC sampling stage and may get majorly removed in the digital filters and down-sampling.
- In theory, it is possible to monitor analog signal strength in the analog baseband, but that would increase the complexity of the analog sections.

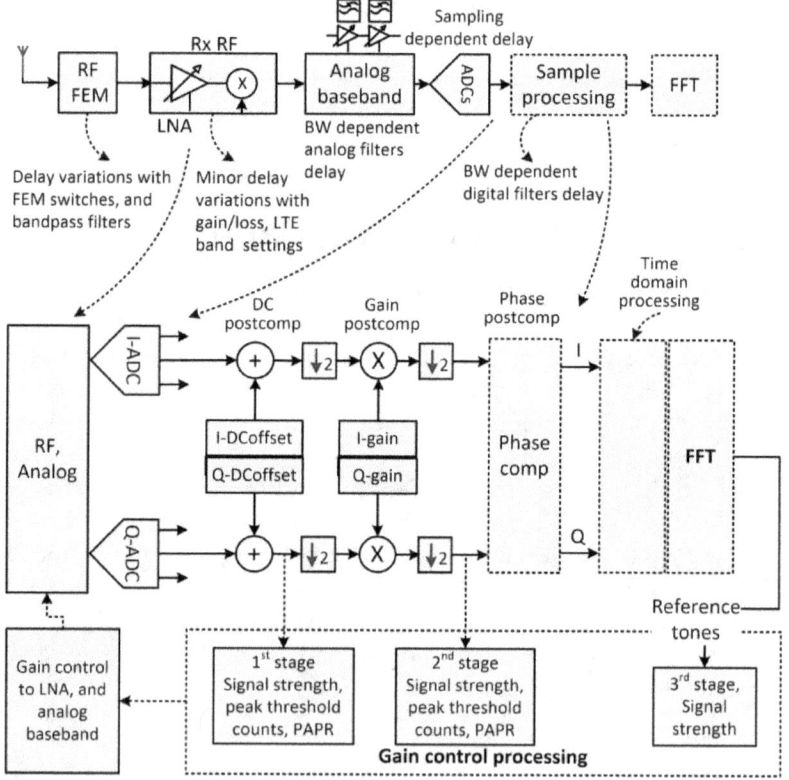

Figure 9.1. Architectural split and functional representation for automatic gain control.

- In practice, the first monitoring of signal strength happens immediately after ADC, preferably after DC offset removal as marked in figure 9.1.
- The signal analysis on digital samples include signal strength estimation similar to root mean square (rms) voltage or receive signal strength indicator (RSSI), peak signal strength counts between two thresholds close to the peak, and peak-to-average-power ratio (PAPR). The first stage analysis immediately on ADC output is a clear representation of the ADC input.
- The next stage of operation is usually after down-sampling operation.
- The third stage is based on frequency-domain reference tones. Here it is more of signal strength without accounting for the interference.
- There could be additional information of RSSI available from frequency-domain processing that can convey information on AGC headroom.
- Referring to the antenna connector point, Rx signals are in the range of −106 to −25 dBm, a dynamic range of 81 dB. RF FEM creates additional 3–4 dB loss. By making use of the ADC dynamic range for weak signals, Rx gain is managed with 70–80 dB, out of which approximately half ~36–40 dB is controlled in LNA and another ~36–40 dB in analog baseband amplifiers.

9.2 Rx Time Alignment

Compared to Tx gain alignment in chapter 8, Rx has simpler timing requirements. Rx is not in direct control of the Tx for timing alignment. As explained in chapter 5, the receiver extracts frequency and time alignment. Once symbol, slot, and frame alignments are extracted, AGC timing alignment is managed with subframe boundary. LNA is a low delay component, and delay alignment for gain control is mainly decided by the analog baseband processing stages. I-Q imbalances are part of digital sample processing. To get better quality in the Rx path specific to gain control, when gain change is significant, and then align the gain changes, DC offsets, and I-Q imbalances with the beginning of the subframe.

9.3 Gain Control Processing

Rx gain control makes use of the information from various stages and estimates the gain gap as excess or deficiency. From the current set gain values, a new gain values are updated, and this becomes a closed-loop control decided by the UE processing.

9.3.1   Signal Strength Estimation

The signal estimation on digital samples can be done in multiple ways, usually implemented with a dedicated hardware operation to handle higher sampling rate samples.

Approach 1. To measure the signal strength, standard deviation (STD)—also the same as rms value—of OFDM signal is used. In most implementations approximations are used that match the results within a fraction of dB, and the error is not significant compared to the signal dynamic range and automatic gain control:

$$s = \sqrt{\frac{\pi}{2}} \frac{1}{N} \sum_{n=0}^{N-1} |x_n| \qquad (9.4)$$

where x_n is the I or Q samples—free from DC offset, and $|x_n|$ is the absolute (abs) of x_n. The signal strength with the sum of abs(I), or abs(Q) can be done independently on I or Q. The results can be combined or either one of them can be used. The ratio of I to Q signal strength estimation can also reveal amplitude imbalance. The sum of the absolute value of samples is multiplied with a constant $\frac{1}{N}\sqrt{\frac{\pi}{2}}$ that equates to STD. This approach allows simplified implementation in hardware or on a processor.

Approach 2. The rms value for N samples of I and Q can be found as $\frac{\sum \sqrt{I^2 + Q^2}}{N}$, and in the literature, such operations are simplified with approximations summarized in [URL (Embedded)], which are well suited for gain control. In one of the approximation, the abs(I) and abs(Q) are taken, and magnitude is calculated as sqrt $(I^2 + Q^2) \approx (15/16)$ Max $(|I|, |Q|) + (15/32)$ Min$(|I|, |Q|)$, where the first coefficient (15/16) is close to 1 and second coefficient (15/32) is close to 0.5. The denominators 16 and 32 are used for easy implementation with mathematical shift operations on digital samples/numbers. In general, the selected coefficients should help easy implementation, and the error magnitude should also get distributed on both sides (positive and negative) from the actual value for the I-Q phase angles from 0 to 90 degrees.

9.3.2   Peak Signal Processing for AGC

As detailed in chapter 3, the OFDM I-Q signals are approximated as the Gaussian probability density function (PDF), thus creating higher PAPR. As a general requirement, the ADC input analog I-Q signal should stay within acceptable (or set) peak clipping. The RF and analog baseband processing can introduce certain nonlinearity for higher peak signal, but may not be clipping. The ADC being a quantizer can clip the signal. Once signal is clipped at ADC (amplitude quantization) to, say, 12-bits, the distortions cannot be recovered with usual signal processing methods.

Hence, it is important to maintain the signal peak level to manageable limits in the AGC closed-loop operation. Signal strength–based operation cannot reveal this peak information. The preferred way for this is to use PAPR estimation and create extra headroom for the required margin.

As illustrated in figure 3.3, the signal level can be monitored for clipping using Gaussian distribution approximations. Taking example of ADC of 12 bits with 1-bit sign and 11-bit magnitude, the extrema (minima and maxima) values are −2,048 and 2,047. Assuming ADC full scale is just occupied with peak level, the desired PAPR arrived from various simulations is 12 dB (a voltage ratio of 4), and the rms level is 2,048/4 = 512. The ADC output can be estimated for rms value, and on the basis of the gap between estimated and desired 512, the gain in RF and analog baseband stages can be adjusted, and the assumption uses PAPR of 12 dB. Often headroom is created to account for short-term signal strength variation due to fading, lag (delay) in gain correction due to response time (or smoothing), and gain settings update errors. The gain update error can happen due to programmed gain value and the actual applied gain in analog and RF stages. When gain change is programmed to 18 dB, but actual gain change may be 18.75 dB, a small error may be present. To avoid AGC steps to keep wandering, a hysteresis is created through the provision of extra headroom. This headroom also helps to deal with process, voltage, temperature (PVT) corners of the various RF FEM, RF, and analog hardware modules.

The peak monitoring helps to enhance AGC operation. The peak sensing can happen on abs (I) or abs(Q). The usual approach is to keep one threshold close to the peak, and observe number samples crossing the threshold. To reinforce further with example, the peak sample value of ADC is 2,047, and 0.5 dB below peak is 1,933. Over a block of N samples, the samples crossing (exceeding) the threshold 1,933 is counted as m_U. If the count m_U/N is less than set threshold—on the basis of the complementary cumulative distribution function (CCDF) as given in chapter 3—then gain has to be increased, and if count m_U/N is more, gain has to be decreased. Over a few blocks of samples or SFs, the gain will settle to the required operating point. If the system design is to create extra 2.5 dB headroom for various margins, the 1,933 threshold will be set lower by 2.5 dB that is 1,450. Refer to chapter 3 for more details on the topics PAPR, crest factors, CCDF, and rms levels. The more advanced version of this would be to keep two thresholds. As continuity of example, threshold 1,450 will be upper limit and another threshold may be set 0.5–1 dB below 1,450. The sample count m_L above the lower threshold, count m_U, and the total number of points N can be used to create arithmetic that will provide more robust estimation tolerating the CCDF fluctuations. On the basis of the mathematical formulation, either the crest factor or PAPR can be used.

9.3.3   Signal Strength from Reference Tones

After frequency-domain (FFT) processing, the tone amplitudes go through channel estimation and finally equalization (chapter 1). The channel equalized tones are used to estimate the reference tone signal power (RSRP), and this RSRP parameters can also be used to help AGC processing. The RSRP is interference free, and it cannot truly represent the signal in the presence of significant undesired signals. The common practice is to start with RSRP (if available), and enhance the signal strength estimation from stage 1 and 2 as per figure 9.1. In general a well-designed PAPR (or crest factor) based system can also work in a close loop, even if signal strength and RSRP are not included in AGC processing.

9.4   DC Offset Issues with AGC

The DC offset removal is crucial in achieving the signal chain full dynamic range, and proper AGC operation. Refer to chapter 6, section 6.2, examples on "DC offset influencing the signal dynamic range" as a background information for this section. As a continuity of section 6.2, DC offset reduces the dynamic range of the ADC. The DC offset creates erroneous STD using simple methods. To be more accurate with STD, variance has to be obtained first for calculating STD. The PAPR-based thresholds explained in previous section with significant DC offset samples will not establish stable estimates and operating points.

In one of the workable options, the estimates for AGC has to be derived on DC offset removed samples, and if the system has to work with DC offset, the offset will be created as headroom in gain that is clearly a loss of ADC dynamic range. The headroom is calculated on the basis of the peak-to-peak (pk-pk) signal level and the DC offset. The ADC with full scale of 1.0 V pk-pk with offset of 0.2 V will limit the ADC usage for 0.6 V pk-pk, a loss of $20 \log_{10}(0.6/1.0) = -4.44$ dB. A 0.6 V pk-pk with 0.2 V offset will create signal going to 0.5 V pk on positive side, and -0.1 V on negative side. Before DC offset building up at ADC input, the offset cancellation in analog sections will be used to reduce the offset at coarse level to a reasonable level of ± 50 mV at ADC input. For ± 50 mV, a loss of $20 \log_{10}(0.9/1.0) = -0.92$ dB headroom is required, which may be acceptable in the system design. In this section and in section 6.2, 1 V pk-pk ADC full scale is taken as an example for easy mathematics. The actual ADCs pk-pk may be of lower voltage, and a suitable arithmetic has to be applied with the pk-pk voltage and DC offsets.

The summary from this section is that the advanced AGC estimation methods should use DC offset–free samples, and uncanceled DC offset will need headroom. Often there will be secondary effect while updating the gain. The gain update causes DC offset to change, and DC offset change is not directly related to gain change as gain change may be distributed across multiple amplifier stages. Hence, it is essential to make sure DC offset estimation and cancellation, signal level estimation, PAPR-based quality metrics/thresholds, and AGC algorithms—including settling time—to work together for achieving optimal dynamic range and stable operating points.

9.5   Blockers

The main undesired signals or conditions that influence receiver sensitivity, selectivity, dynamic range, AGC, nonlinearity, and distortions are as follows:

- Wide-band signals (BW up to 5 MHz) used with adjacent channel selectivity.
- In-band blockers with wide-band (BW up to 5 MHz) signals.
- Out-of-band (OOB) blockers with continuous wave (CW) tone.
- Narrow-band blockers with CW tone.
- Tx leakage from RF FEM, which is not a blocker by classification.

9.5.1   Desensitivity

In this section word "desense" is used to convey desensitivity. An interference entering the receiver distorts or creates additional noise/interference components that degrade the final operating point for 95% of maximum LTE throughput. For example, consider that the Rx is operating under normal conditions with receiver sensitivity (REFSENS) of, say, −98 dBm to achieve 95% of the maximum throughput. With interference, the throughput is below 95%. The usual way is to add extra margin (increase in signal power) on top of REFSENS to achieve 95%. To continue with the example, the new level (with the presence of interference) to achieve 95% is −95 dBm, an extra 3 dB power. This conveys that the interferer is desensitizing the capability of the receiver by 3 dB and is simply referred to as desense of 3 dB.

9.5.2   Adjacent Channel Selectivity (ACS)

As shown in figure 9.2, the adjacent channel interference is present next to the desired channel with a frequency gap of only a few kilohertz. The receiver's capability to reject or work with adjacent channels is checked with reference to 95% of the maximum throughput. Figure 9.2 is shown for LTE BWs 1.4, 5, 10, and 20 MHz, and other BWs 3 and 15 MHz [3GPP 36.521-1] are not shown and they follow similar BW-dependent parameters. The summary on ACS is listed as follows:

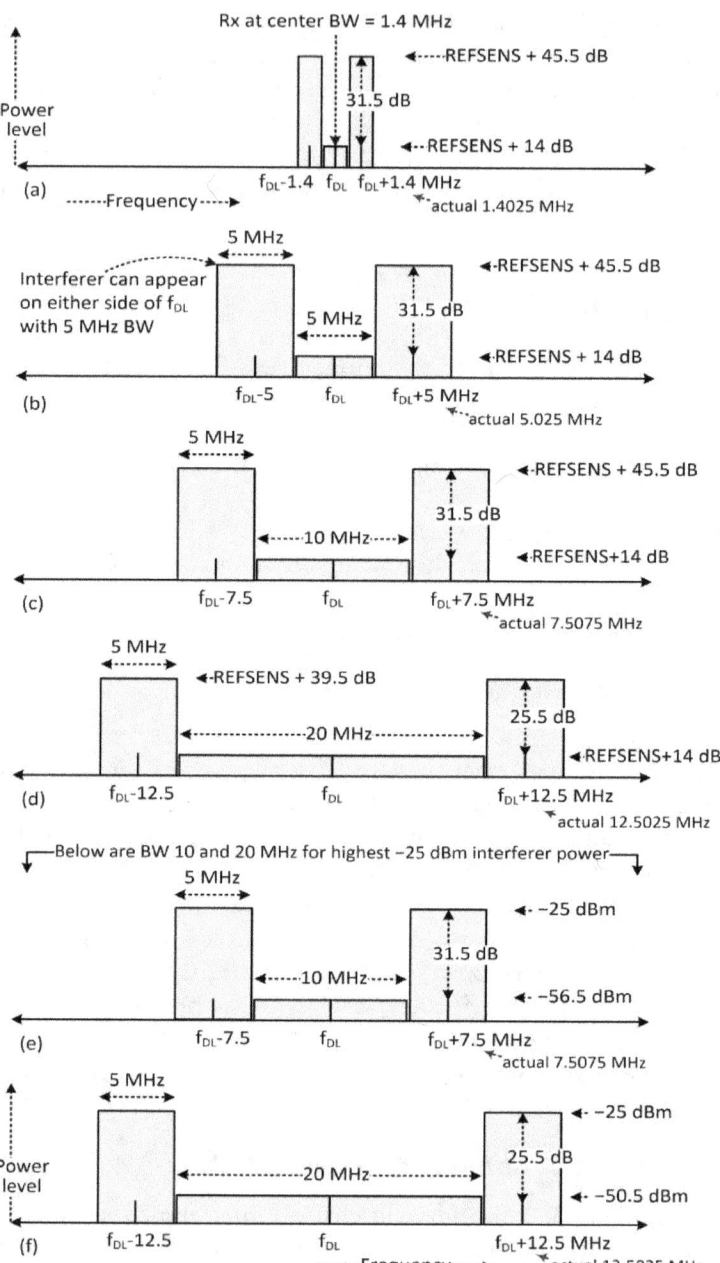

Figure 9.2 Adjacent channel selectivity [3GPP 36.101, 36.521-1]. (a) BW 1.4 MHz case 1. (b) BW 5 MHz case 1. (c) BW 10 MHz case 1. (d) BW 20 MHz case 1. (e) 10 MHz case 2. (f) 20 MHz case 2. Refer to 3GPP 36.521-1 for case 2 description.

- The undesired signal is centered at offsets 1.4, 5, 7.5, and 12.5 MHz respectively for LTE BWs 1.4, 5, 10, and 20 MHz. The frequency offsets are given up to fourth decimal values in 3GPP specifications, and in the figures and descriptions nearest decimal values are used.
- Considering LTE 10 MHz BW, occupied BW is 9 MHz, and this provides spacing of (7.5075 − 2.5 − 9/2 =) 0.5075 MHz between band edge of occupied and the starting edge of undesired ACS test signal.
- If LTE channel BW is only 10 MHz (similar to Band 13 and Band 17), the duplexer uses ~10 MHz filters and ACS undesired signal may get removed significantly. In case of wider LTE band, examples as in Band 2 with band span of 60 MHz (that is much wider than LTE 20 MHz channel BW), the duplexer or SAW filters in RF FEM may not be helping to remove undesired frequency components.
- Referring to chapter 2 and figure 9.1, the undesired signal passes through RF FEM, LNAs, and down-conversion mixer. Signal nonlinearity in case of −25 dBm desired signal has to be considered for meeting the specifications.
- The analog filters after mixer stage can keep attenuating and ADCs can see part of ACS signal, and additional undesired signal residues can be removed in digital filtering. The frequency-domain processing keeps the undesired tones away from band edge, but this can cause ICI at band edges—if ACS occupies significant amplitude.
- In general ACS signal need not be present with tones orthogonal to desired signal tones; hence, the ACS undesired signal may not align with 15 kHz tone spacing in symbol duration. In theory, the residual signal can appear in the side lobes of the sinc functions. For improving the quality, the influence of ACS has to be taken care across multiple stages of RF, analog, and digital filters without allowing it to get into FFT input.

9.5.3   In-Band Blockers

As represented in figures 9.3 and 9.4, in-band blockers are wide-band signals (similar to LTE signals). The in-band has four cases [3GPP 36.521-1], and in figure 9.3 case 1 is illustrated with the following summary:

- For Rx DL 1.4 MHz BW, the interferer BW is 1.4 MHz and can reside on either side. Taking example of REFSENS as −98 dBm, and desensitivity (desense) of 6 dB, the Rx signal in the in-band is −98 + 6 = −92 dBm. A strong desired signal exceeding REFSENS by 6 dB is allowed when blockers are present.
- For 3 MHz—not shown in figure 9.3—the interferer BW is 3 MHz.
- For 5 and above MHz, the interferer BW is 5 MHz. With wider Rx

BW, the interferer position is further away from the center.
- For 20 MHz the REFSENS is more desensitized compared to other BWs.

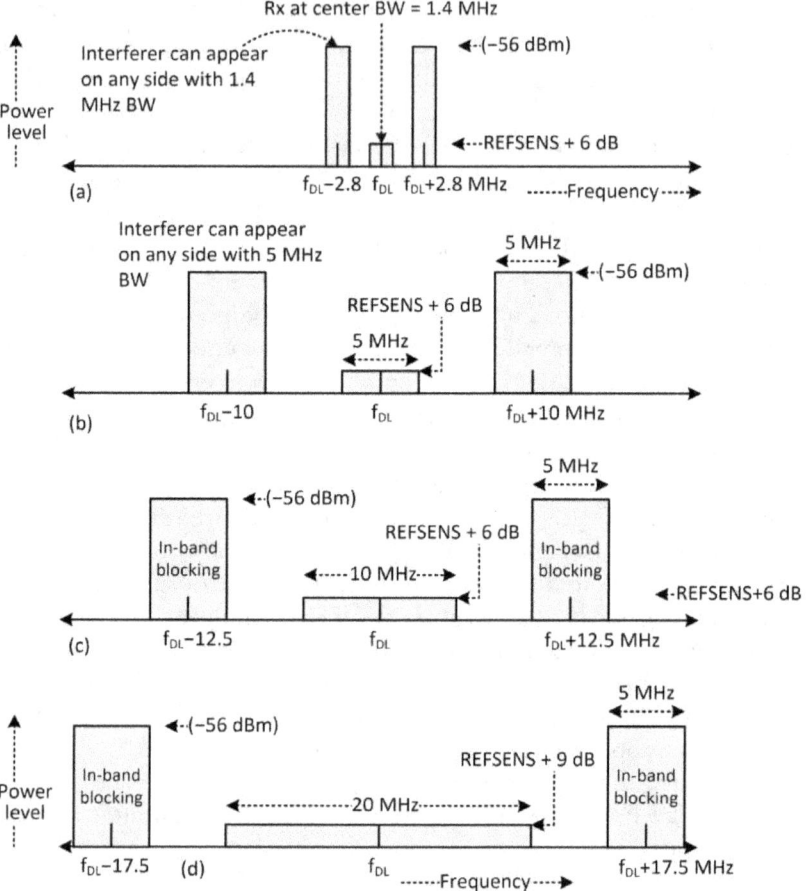

Figure 9.3. In-band blocker shown for LTE BW 1.4, 5, 10, and 20 MHz for interferer offset case 1 [3GPP 36.101, 36.521-1].

As shown in figure 9.4 for in-band blockers of case 2, the blocker strength is higher −44 dBm for case 2 versus −56 dBm for case 1, and blocker location is further away from the center frequency compared to case 1. The in-band blocker influences the signal dynamic range, AGC, nonlinearity, and distortions. Some LTE bands, such as Bands 13 and 17, are of maximum 10 MHz BW. For LTE maximum BW of 10 MHz, the duplexer with ~10 MHz BW may not be helping 1.4 and 3 MHz as in-band blockers for low BW fall inside the duplexer BW.

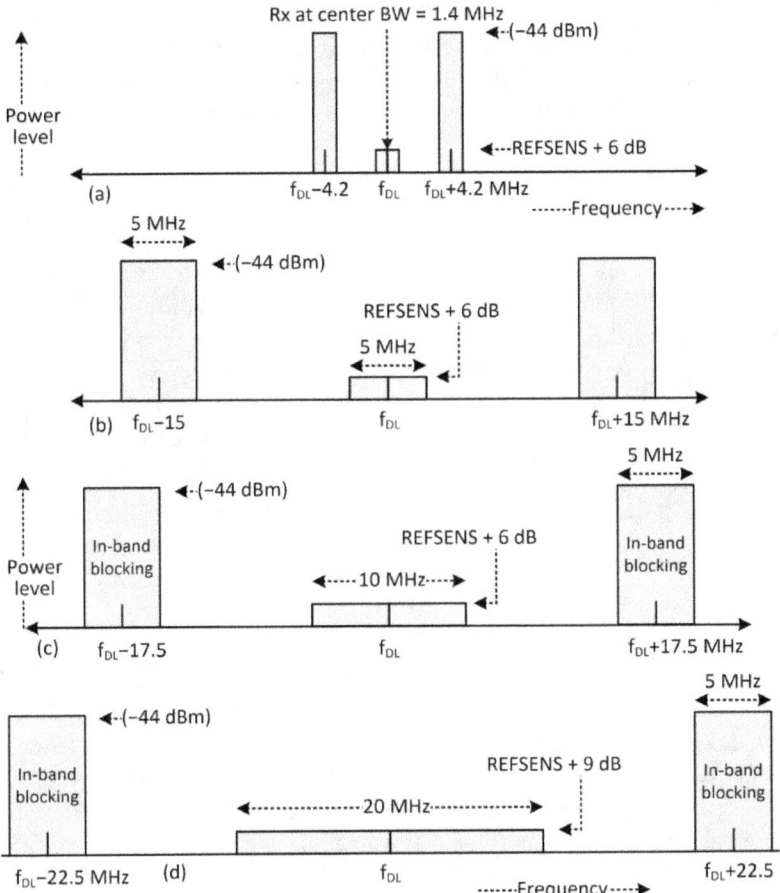

Figure 9.4. In-band blocker shown for LTE BW 1.4, 5, 10 and 20 MHz for interferer offset case 2 [3GPP 36.101, 36.521-1].

9.5.4   Out-of-Band (OOB) Blockers

The out-of-band blocker is a strong continuous wave (CW) tone that can appear on either side of the band edge starting from $f_{DL}+15$ MHz or $f_{DL}-15$ MHz. With the gap from the band edge, the power level of CW tone also increases—meaning strong interference with increased frequency from LTE carrier center frequency. In figure 9.5, the tone envelope is shown to appear starting from 15 MHz away from f_{DL}, and *it is actually one CW tone that can appear anywhere in the marked limits and at the specified power level.* The OOB majorly gets attenuated in the RF FEM surface acoustic wave (SAW) filters or duplexers, and the selection of such filters can influence the rejection level. The desense for OOB is 6–9 dB depending on the LTE BW [3GPP

36.521-1]. Figure 9.5 is shown for 20 MHz, and it is the same representation for other LTE BWs with LTE signal at f_{DL} and OOB appearing starting at 15 MHz away from f_{DL}. Below 20 MHz, additional markings are made for 10 MHz and 5 MHz with a different desense level.

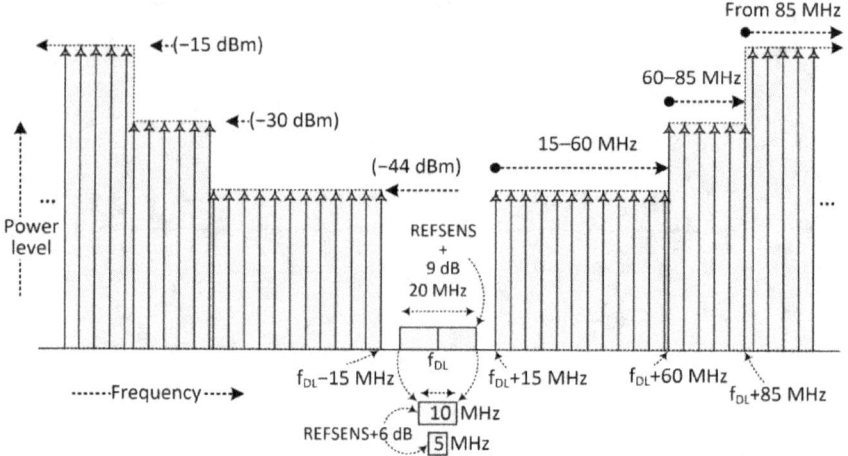

Figure 9.5. Out-of-band blocker shown for LTE BW 20 MHz [3GPP 36.521-1]. The OOB is shown for a wide range of frequencies on both sides of the LTE channel. The OOB is one CW tone that can present in the marked frequency span with the specified CW power level envelope. The OOB is the same for other BWs (1.4, 3, 5, 10, and 15 MHz) with the change of the desense value [3GPP 36.521-1], given as 7 dB for 15 MHz, and all other BWs 10 MHz and lower as 6 dB.

9.5.5   Narrow-Band Blockers

Narrow-band blocking is a strong CW tone at −55 dBm immediately after LTE band. As marked in figure 9.6, the frequency gap from LTE channel BW edge is approximately 200 kHz. As LTE-occupied BW tones are not present up to full band, the gap from occupied BW (with actual OFDM tones) is more than 200 kHz. For 1.4 MHz, the useful BW is 1.08 MHz, and frequency gap is 367 kHz. For 10 MHz, the OFDM tones span for 4.5 MHz on either side or the frequency gap is 500+200 = 700 kHz. At higher BW the separation of interferer with the occupied band edge is higher, and for 20 MHz the separation is 1.2 MHz. Narrow-band blockers can majorly go through the RF FEM and may start getting attenuated only after down-conversion mixer. The CW tone along with LTE in-band blockers can create intermodulation products that interfere with the desired signal. As the attenuated version of narrow-band blocker appears at the ADC input, the preceding stages have to account for enough linearity and the parameters of such requirements can be arrived through system simulations.

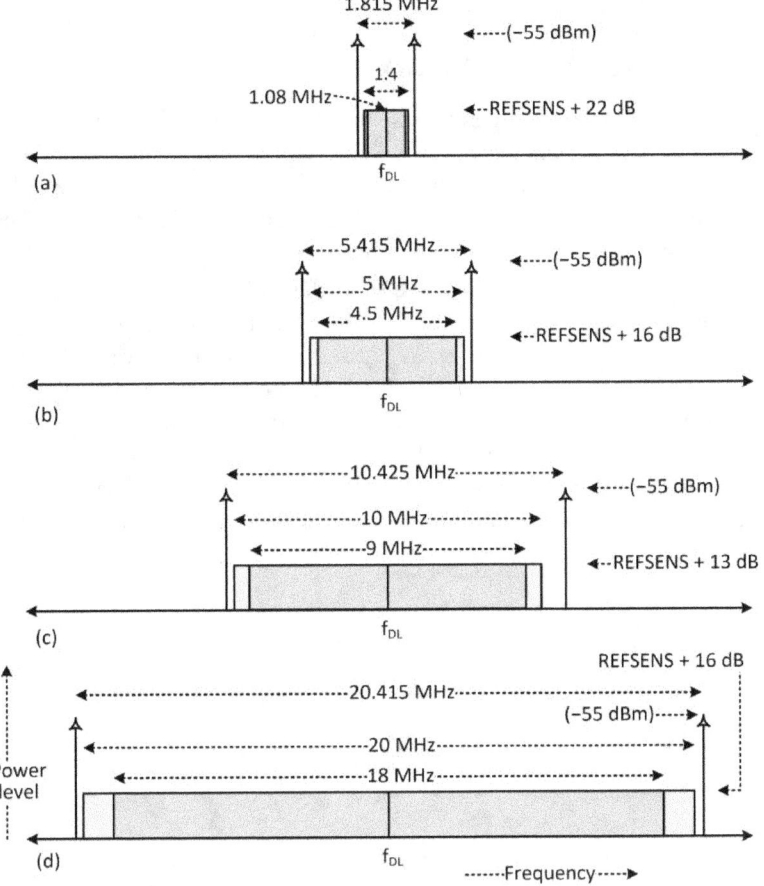

Figure 9.6. Narrow-band (CW) blocker for LTE BW [3GPP 36.521-1]. (a) 1.4 MHz. (b) 5 MHz. (c) 10 MHz. (d) 20 MHz.

9.5.6   Blockers and AGC

Multiple signal-quality aspects have to be considered with blockers as summarized below:

- The blocker signal may partially get rejected in the signal chain and a significant part of the blocker signal can remain even at ADC input.
- The signal rms value and PAPR may not be following usual relations with blockers. Hence, it is essential to activate all the three main things (see section 9.3), such as signal strength across various stages, PAPR analysis, and RSRP from frequency-domain processing.
- The signal strength across multiple stages shown in figure 9.1 should

also derive information to fine-tune analog baseband filters and digital filters to help minimize distortions through blockers.

9.6   Chapter Summary

The blockers play a significant role in the receiver performance. As the blockers are of higher than the desired signal amplitude, the RF FEM, LNA, and mixer stages should possess required intermodulation characteristics, which are not covered in this book. The REFSENS with desense margin is for quadrature phase shift keying (QPSK), where the expected signal to noise ratio (SNR) for demodulation and detection is ~0 dB (varies by system design). Hence, the tests will be a little bit more tolerant. In an actual scenario with higher QAM modulations, the SNR (or Rx power) will be higher, and interference may get merged with the noise. Specific to this chapter, the overall performance is significantly dependent on hardware and firmware combination in terms of frequency response, dynamic range, intermodulation, DC offsets, adaptive characteristics of AGC, and capability to sense and fine-tune the frequency responses, and non-linearity.

10 SIGNAL-QUALITY CHARACTERISTICS AND MEASUREMENTS

The chapter is oriented to user equipment (UE) signal quality. UE quality can also be observed through transmit (Tx) and receive (Rx) antenna ports. The Tx on UE (up to release 8/9) is assumed to keep one antenna connector, and Rx should have two antenna connectors—one for the main and the other for diversity (chapter 2). The Tx and Rx quality requirements are spread across 3GPP 36.521-1 and 36.101. The base features are in 3GPP release-8 documents, and expanded text is from release-9 documents and later versions. For first-level understanding and introductory material, refer to [URL (Aeroflex-7100), URL (R&S-1), URL (R&S-2), URL (NI-testing)].

10.1 Transmitter Characteristics

The transmitter characteristics in 3GPP 36.521-1 are grouped as follows [3GPP 36.521-1, Christina Gebner]:

- Transmit power for maximum UE output power, maximum power reduction (MPR), additional MPR, and configured UE transmitted power.
- Output power dynamics for minimum output power, transmit OFF power, ON/OFF time mask, and power control.
- Transmit signal quality for frequency error, error vector magnitude (EVM), local oscillator (LO) leakage, in-band emission, and spectrum flatness. Output RF spectrum emissions for occupied BW, out-of-band (OOB) emissions, and spurious emission.
- Transmit intermodulation with continuous wave (CW) and modulated signals.

Table 10.1. UE transmitter tests with reference to 3GPP 36.521-1. In the table 'min' is minimum and 'max' is maximum. The typical values in the table will have tolerance limits, refer [3GPP 36.521-1] for more details.

3GPP 36.521-1 section		Section name	Short description	Typical value
6		Transmitter characteristics	Specified at the UE antenna connector with a single Tx antenna	
6.2		Transmit power	Check configured max power and network signaling (NS) back-off	
	6.2.2	UE maximum output power	UE max power is within acceptable limits	23 ± 2 dBm
	6.2.3	Maximum power reduction	Lower side tolerance on max output power	1 to 2 dB
	6.2.4	Additional maximum power reduction (A-MPR)	Network signal (NS) specific power reduction to help ACLR & emissions	1–3 dB as per NS_number
	6.2.5	Configured UE transmitted output power	To check UE is working with min of two {maximum allowed power set by eNodeB, and power set by UE power class} levels	Class-3 uses 23 dBm
6.3		Output power dynamics	Checks min, OFF power, OFF/ON timings, ramp-up, ramp-down in multiple steps	−40 to 23 dBm
	6.3.2	Minimum output power	Check UEs capability to transmit min power	−40 dBm
	6.3.3	Transmit OFF power	Tx OFF power in non-transmission or blanking	−50 dBm
	6.3.4	ON/OFF time mask	Ramping time allowed for Tx ON to OFF, Tx OFF to ON. Varies from 20 to 40 μs based on scheduled symbols in a subframe	20 μs
	6.3.5	Power control	Power ramp-up and ramp-down steps, and tolerances in multiple power control configurations	
	6.3.5.1	Power control absolute power tolerance	UE Tx is able to start at a specific power after a gap larger than 20 ms	±9 dB
	6.3.5.2	Power control relative power tolerance	UE Tx is able to start at a specific power relative to previous subframe of gap ≤20 ms	±2.5 dB for <2 dB power step
	6.3.5.3	Power control aggregate power tolerance	In non-contiguous transmission with gap within 21 ms, and zero dB power change; Tx to maintain power within tolerance limits	±2.5 dB PUCCH, ±3.5 dB PUSCH
6.5	6.5	Transmit signal quality		
	6.5.1	Frequency error	UE Tx output error is less than ±(0.1 ppm + 15 Hz) with reference to Rx input	±(0.1 ppm + 15 Hz) over one slot

	6.5.2	Transmit modulation		
	6.5.2.1	Error vector magnitude (EVM)	Modulation EVM to achieve better than QPSK 17.5%; 16-QAM 12.5%; 64-QAM 8%	
	6.5.2.2	Carrier leakage	LO leakage at Tx center frequency	
	6.5.2.3	In-band emissions for non-allocated RB	I-Q imbalances, acceptable level varies based on the tones location	
	6.5.2.4	EVM equalizer spectrum flatness	Spectral variation (ripple) across Tx tones	
6.6		**Output RF spectrum emissions**		
	6.6.1	Occupied bandwidth	Min BW containing 99% of total energy 99% BW is ≤1.4 MHz in LTE 1.4 MHz channel 99% BW is ≤20 MHz in LTE 20 MHz channel	
	6.6.2	Out-of-band (OOB) emission	Unwanted emissions outside LTE channel	
	6.6.2.1	Spectrum emission mask	Tx emissions does not exceed the set limits	
	6.6.2.2	Additional spectrum emission mask	Additional requirements valid in some bands, signaled by network as per NS value	
	6.6.2.3	Adjacent channel leakage power ratio	ACLR levels in adjacent UTRA, and E-UTRA (LTE)	
	6.6.3	Spurious emissions	Unwanted emissions such as harmonic and parasitic emissions, inter-modulations and frequency conversion products. Spurious emissions are next to OOB in frequency	−36 to −30 dBm, [ITU-R SM.329-10 (02/2003)]
	6.6.3.1	Transmitter spurious emissions	To verify UE Tx does not cause interference to other channels	−36 to −30 dBm
	6.6.3.2	Spurious emission band UE co-existence	To verify UE Tx does not cause interference to other bands	
	6.6.3.2A	Spurious emission band UE co-existence (release 9 and forward)	To verify Tx does not cause interference to other bands, additional notes in release-9	
	6.6.3.3	Additional spurious emissions	To impose additional network signaling requirements with spurious emissions	as per NS
6.7		**Transmit intermodulation**	Tx inter-modulations created from wanted signals, and external interferences (−40 dBc) received through Tx antenna	−35 to −29 dBc

10.2   Receiver Characteristics

The receiver characteristics in 3GPP 36.521-1 are for characterizing receiver sensitivity (minimum signal for 95% of maximum throughput), the maximum input signal level, adjacent channel selectivity to receive desired signal with interference, various in-band and out-of-band interferences, narrow-band blockers, wide-band intermodulation, and spurious emissions.

Table 10.2. UE receiver tests as per 3GPP 36.521-1. In the table 'min' is minimum and 'max' is maximum. The typical values in the table will have tolerance limits, refer [3GPP 36.521-1] for more details.

Section/ sub-section		Section name	Short description	Typical value
7		**Receiver characteristics**	**Specified at the UE antenna connector**	
7.1		General	Rx characteristics are specified at antenna connector(s) of UE with default signaling value NS_01. For UEs with more than one antenna, identical interfering signals shall be applied to each antenna connector.	
7.2		Diversity characteristics	LTE UE receivers are made with two receiver ports as baseline and both the ports have to be verified for simultaneous operation.	
7.3		Reference signal level ($P_{RFESENS}$)	Minimum Rx signal level that gives 95% of max throughput. Observed in QPSK. In FDD Rx sensitivity is observed during Tx transmissions. The sensitivity levels vary by LTE band, and channel BW.	−106 to −91 dBm
7.4		Maximum input level	UE capability to work at maximum Rx power. This helps checking UE capability for close-by range from eNodeB.	−25 dBm
7.5		Adjacent channel selectivity (ACS)	To check capability to receive desired signal in the presence of adjacent channel given as ratio of desired to adjacent channel response.	27–33 dB
7.6		Blocking characteristics	Rx ability to receive desired signal in the presence of interferers (blockers).	
	7.6.1	In-band blocking	1.4 to 5 MHz BW interfering signal is at 2.1 to 12.5 MHz (less than 15 MHz) from LTE desired signal on either side of band-edge.	Multiple values
	7.6.2	Out-of-band blocking	Strong CW interferer falls outside of LTE desired signal by more than 15 MHz. CW power increases up to −15 dBm with increased frequency spacing from desired LTE channel. REFSENS power is adjusted based on BW.	Multiple values
	7.6.3	Narrow-band blocking	Ability to maintain >95% of max throughput in the presence of CW interferer adjacent to desired LTE channel. Additional margin is provisioned on top of REFSENS power.	CW at −55 dBm
7.7		Spurious response	Extension of out-of-band blocking to check response at multiple interferer frequencies.	

7.8		Intermodulation characteristics	To validate intermodulations (IM), two or more interfering signals are used with specific relationship to the desired signal.	
	7.8.1	Wide-band intermodulation	A CW and modulated interferers at −46 dBm are used to test IM.	
7.9		Spurious emissions	Receiver's radiated emission in the frequency range 30 MHz to 1 GHz (measurement BW 100 kHz), and 1 GHz to 12.75 GHz (BW 1 MHz) range.	

10.3   EVM and SNR

EVM is the error vector magnitude—specified commonly as percentage (also in dB units). Table 10.3 lists the EVM percentage, dB units, and signal-to-noise ratio (SNR) relations. As stated in [URL (Keysight-EVM)] EVM is defined in multiple ways. Specific to LTE QPSK (quadrature phase shift keying), 16-QAM (quadrature amplitude modulation), and 64-QAM, the ideal constellation points provide the reference signal, and the actual constellation points are with noise and distortions. The difference between the actual received or transmitted signal constellation points to ideal reference is the error vector. The error vectors are used for computing the rms value and normalized to the ideal constellation point. When error is referred to as noise created on the rms signal (in this case ideal reference constellation), the SNR and EVM are related as $SNR\_dB = -20 \log_{10}(EVM\%/100)$ and EVM Percentage $= 100 \times 10^{\wedge}(-SNR\_dB/20)$. The table lists the SNR and EVM examples for multiple values. When specifying the amplifiers quality, the EVM is also specified in −dB units. The EVM percentage and EVM in dB scale are related as EVM in dB $= 20 \log_{10}(EVM\%/100) = -SNR\_dB$. In generic discussions, EVM may be simply stated as, say, 40 dB; probably the meaning is −40 dB, and EVM is 1%, and SNR is 40 dB.

As marked in the table notes, the 3GPP 36.101 has recommended EVM limits for LTE Tx as 17.5% for QPSK, 12.5% for 16-QAM, and 8% for 64-QAM. The lower EVM percentage conveys better quality. With improvements in RF semiconductor devices, the EVM percentages are much better (lower EVM% values) in practical Tx. As LTE UE Tx operates from −40 to +23 dBm, the EVM values vary as the power level changes. As marked in the table, the EVM values are approximately in the range of 2.5%–3.5% for very low power (−40 to −30 dBm) and at higher power (+18 to +23 dBm). At lower power various noise components, and distortions contribute to increase EVM percentage. At high power PA nonlinearity and FEM components may limit the EVM. At midrange the EVM will be usually better.

Table 10.3 EVM and SNR relations

EVM Percentage = $100 \times 10^{(-SNR\_dB/20)}$;				
SNR_dB = $-20 \log_{10}(EVM\%/100)$;				
EVM dB = $20 \log_{10}(EVM\%/100) = -SNR\_dB$				
EVM%	EVM dB	SNR dB = $-20 \log_{10}(EVM\%/100)$	LTE Tx quality notes
0.5	−46.0	46.0	<0.5% is instruments quality
1.0	−40.0	40.0	
1.5	−36.5	36.5	Low (means better) EVM at mid-range Tx power operating EVM
1.8	−34.9	34.9	
2.0	−34.0	34.0	
2.2	−33.2	33.2	
2.4	−32.4	32.4	
2.6	−31.7	31.7	
2.8	−31.1	31.1	Higher (less quality) EVM at low Tx power and high Tx power
3.0	−30.5	30.5	
3.2	−29.9	29.9	
3.4	−29.4	29.4	
3.6	−28.9	28.9	
3.8	−28.4	28.4	
4.0	−28.0	28.0	
6.0	−24.4	24.4	
8.0	**−21.9**	**21.9**	3GPP 8% max EVM limit for 64-QAM
10.0	−20.0	20.0	
12.0	−18.4	18.4	
12.5	**−18.1**	**18.1**	3GPP 12.5% max EVM limit for 16-QAM
14.0	−17.1	17.1	
16.0	−15.9	15.9	
17.5	**−15.1**	**15.1**	3GPP 17.5% max EVM limit for QPSK
20.0	−14.0	14.0	
31.6	−10.0	10.0	

Notes: EVM is shown as related to SNR in table 10.3. Use table 10.3 for first order approximations. Actually EVM is complex math, and depends on many other signal quality impairments such as frequency errors, phase-noise/jitter, and I-Q imbalances.

10.4   Tx ON/OFF Mask and Transient Time

Various ON/OFF masks are provided in 3GPP 36.101 section 6.3.4. The ON means that UE Tx is sending the required and controlled power level from −40 to +23 dBm (referred to antenna port) with some deviations depending on network signaling and permitted tolerances. The OFF power of Tx is −50 dBm or lower. In the worst case, Tx will operate at +23 dBm in ON, −50 dBm in OFF, and the sudden change of power may be 73 dB; this takes finite time referred to here as the transient period. The power levels referred to in this section are from the UE Tx antenna port. The 3GPP 36.101, section 6.3.4, has multiple scenarios with illustrations for physical uplink shared channel (PUSCH), physical random access channel (PRACH), sounding reference signal (SRS) transient time boundaries at symbol, slot, and subframe (SF). The 3GPP 36.101 specification allows ~20 μs (microseconds) for the general ON/OFF time mask transient period, which is a relaxed specification for the current technology.

As explained in chapter 8, Tx gain control, the power control and power blanking are required to be implemented at symbol boundary—meaning ON/OFF mask has to work at symbol boundary. The transient period of 20 μs may start differently for different scenarios [3GPP 36.101]. The converging point for attention is the period 20 μs is a higher number, and the implementations have to work in less than cyclic prefix (CP) duration ~5 μs, which is possible to achieve with the current technology of RFIC, FEM, analog processing, and digital processing. Taking the example of power change at subframe n and $n+1$ boundary, the subframe n is fully protected without any overlap of transient period, and transient period starts at the beginning of the subframe $n+1$. The 20 μs transient time can distort CP, and also the first symbol of SF($n+1$) will be windowed on the basis of the transient envelope and phase characteristics of transient response. The orthogonality of the tones in the first symbol can get distorted as the full symbol duration does not contain useful signal. To optimize on quality, the transient time has to be controlled to less than CP duration when transient is falling on active (nonblanking) symbol. The techniques highlighted in chapter 8 for aligning from digital samples to RF PA gain change have to be properly used to improve signal quality.

10.4.1   ACLR

ACLR is the adjacent channel leakage power ratio, and is a measure of the amount of power leaking into adjacent channels. ACLR levels appear in adjacent UTRA, and E-UTRA (LTE). Refer to [3GPP 36.521-1, Christina Gebner, URL (NI-PA)] for more information on this topic. ACLR is the ratio of the filtered mean power on the assigned channel to the filtered mean power in the adjacent channel. When PA is used in multiple modes with envelope tracking (ET), the ACLR becomes the dominant parameter of distortions.

10.5   Signal-Quality Connections Summary

A summary of LTE tests as per 3GPP 36.521-1, sections 6 and 7, are listed in tables 10.1 and 10.2. These are more close to the transmission characteristics and the performance under various constraints to match the usage environment. In this section, relevance with tables 10.1 and 10.2 and the signal quality and analysis are presented. At high-level the signal quality gets influenced as summarized below:

- The parameters amplitude, phase, time, frequency, and spatial characteristics (coupling, propagation, and fading) and their variants influence the signal quality.
- The hardware and firmware controlling the parameters, their adaptability under various electrical, and environmental conditions help change the dynamics of signal quality.
- Once hardware is involved with RF and analog, the process variations, voltage, temperature (PVT), and mechanical vibrations and shocks influence the quality.
- The antenna orientation, patterns, efficiency (gain and losses), directivity, incorporation of antenna and RF FEM tuning for improved matching, active antenna provision, and electromagnetic environment UE is operating all influence the signal quality.
- When multiple wireless devices are coexisting in the proximity, the wireless device protocols that share the time and frequency grid also significantly influence the signal quality.

10.6   Signal-Quality Measurements

LTE testing uses several families of instruments, and this chapter is focused on UE specific instruments. In a typical usage, eNodeB is the master and UE is the slave device. It is a closed-loop operation mainly decided by the eNodeB. In the simplified interpretation, eNodeB is a signal source for data network with LTE signal generation and LTE signal analyzer. The analyzer and generator have to follow LTE 3GPP protocols and such operation in an instrument is called LTE call generation.

10.6.1   LTE Testing with Call Generators

Figure 10.1 is a LTE call generator that works in a controlled way. As basic operation, the UE gets attached to eNodeB (instrument), and continues UL/DL transmissions. There are modes in the instrument for low-level testing mainly for RF and physical layer that need configurations in UE and eNodeB (test instrument is eNodeB here). Taking example of URL (Aeroflex), the instrument establishes the UE attach and continues data transmission and conducts LTE 3GPP 521-1 section 6 Tx and section 7 Rx measurements. Similarly R&S [URL (R&S-1), URL (R&S-2), URL (R&S-LTE)] has low-level application programming interface (LLAPI) test suit that can perform 36.521-1 section 6 & 7 tests. The medium level API (MLAPI) package [URL (R&S-CMW500)] includes the signaling tests. The Anritsu [URL (Anritsu-8821C), URL (Anritsu-measurements), URL (Anritsu-8820C)] has a wide collection of test instruments, and they can be used for 36.521-1 measurements. In general, the instruments come with several firmware/modules, and the actual capability depends on the hardware and firmware modules resident in the instruments.

Figure 10.1(a) denotes a call generator instrument working as scaled down version of eNodeB. The RF signal appears either as combined UL + DL or separate UL and DL on RF connectors. The attenuators, splitters, and circulators are shown to represent for proper splitting and RF distribution of the signal:

- The (programmable) attenuator controls attenuation in the path without keep using instrument attenuation controls.
- The splitters divide equal power to two or more ports.
- The circulator usually will have three ports. Assuming it is connected to UE-Rx, UE-Tx, and antenna (goes to eNodeB as instrument). The Tx signal from UE will go to antenna, and the received signal from antenna will go to the Rx port. In figure 10.1(a), instrument DL goes to UE, and UE UL goes to instrument.

Figure 10.1(b) is an extension of figure 10.1(a) with the inclusion of internal or external fader [URL (Azimuth)] with noise source. Using fader various channel models can be included in the DL. Figure 10.1(c) is an extension of figure 10.1(b) with the inclusion of extra air interface monitor, and for this instrument from Sanjole [URL (Sanjole)] is taken for example. The instrument can monitor signal on air interface or on RF cables (from splitter signal), and analyzes both DL/UL with call progress states and messages. The instrument can look at very low-level signal modulations in addition to messages and signaling.

In the top path of figure 10.1(a)-(c), signal synthesizer (10–26 MHz) is shown that takes 10 MHz from the instrument reference outputs [URL (Keysight-MXG), URL (litepoint1)] and delivers frequency locked 26 MHz output to UE. The instruments are made with high-precision OCXOs, and refer to chapter 4 for an overview on frequency references. The UE in its default condition operates with more ppm clock error. For establishing the initial benchmarks of UE performance, it is highly recommended to provision frequency locked instrument clock that eliminates frequency errors and provides best possible LTE reference quality or capability of UE. In the example description here, UE is assumed to be provisioned for external clock in the initial quality assessment.

> Notes: In the figures, the synthesizers are shown to convert 10 MHz to 26 MHz reference. The synthesizers have to maintain better phase and frequency characteristics than the UE local frequency references—such as crystal oscillators or TCXOs. This is especially important to verify as some of the RF synthesizers (instruments) when used at lower (26 MHz) frequency may perform inferior than TCXO phase noise characteristics.

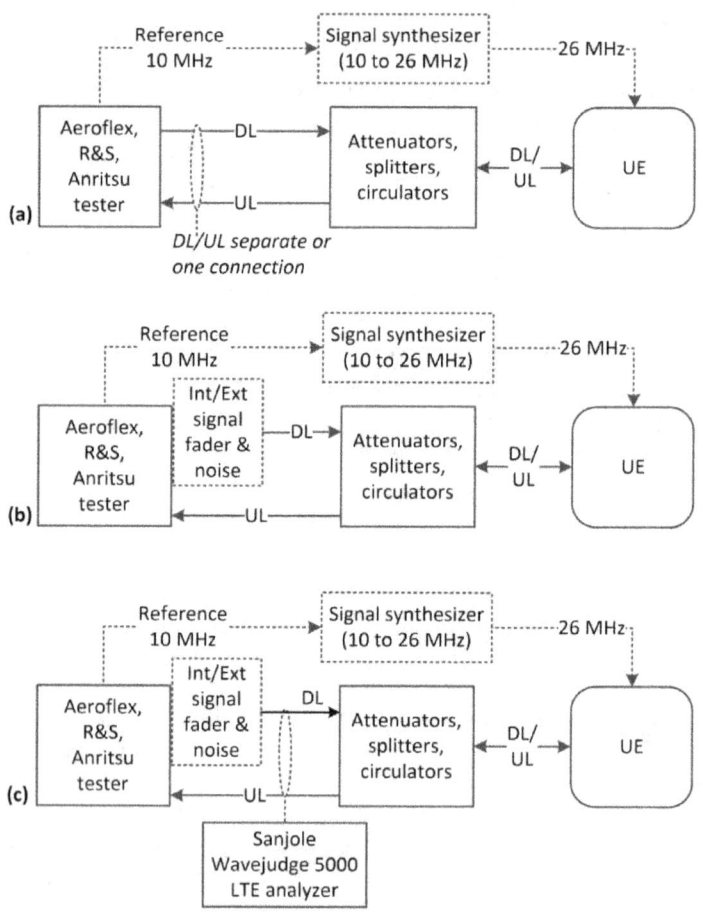

Figure 10.1. UE or modem testing with LTE call generator. (a) UE making use of reference. (b) The DL is going through fader and noise simulations. (c) The DL and UL are intercepted and analyzed through air interface LTE analyzer.

10.6.2   Low-Level Measurements

In the setup shown in figure 10.1, the main focus was to characterize UE Tx and Rx starting from high-level call-flow procedures. For microlevel issues a different setup is preferred that sends the physical-layer signals in a controlled way. The MXG and PXA [URL (Keysight-LTE), URL (Keysight-Analysis), URL (Keysight-PXA), URL (Keysight-MXG)] setup is shown as an example in figure 10.2(a) that can use firmware [URL (Keysight-VSA)] on the instruments to access signal quality. The setup can be further enhanced with internal or external faders and noise generators.

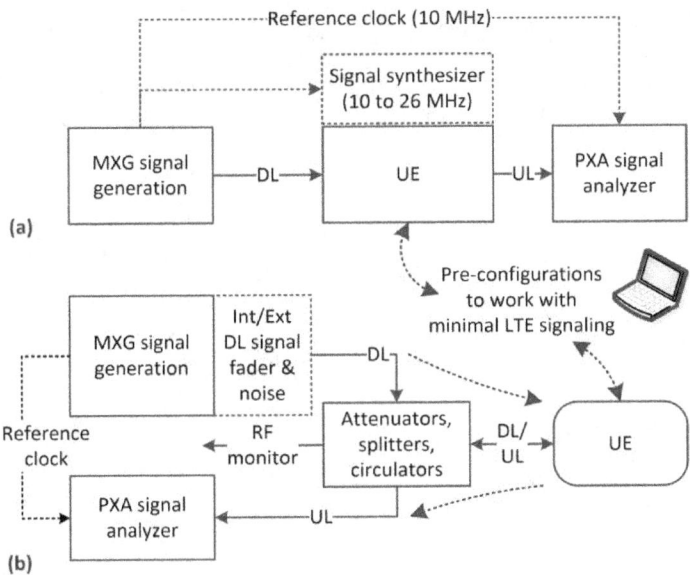

Figure 10.2. UE or modem testing with signal generation and signal analyzer. (a) Basic test setup that also used reference from the signal generation. (b) Extended setup with internal or external fader and noise generation.

10.6.3   Typical Production Setup

In the production tests, a few selected tests are conducted on the basis of the product requirements that checks both DL and UL. The purpose is to activate the basic functions of FEM, RFIC, BBIC, and any associated application processors for a brief amount of time. For this approach, the instruments [URL (litepoint2), URL (NI-testing)] will be directly interfacing with the UE and instrument sends various configurations, and RF signals to the UE in time synchronism, and the UE transmissions are measured in closed-loop interactions (figure 10.3).

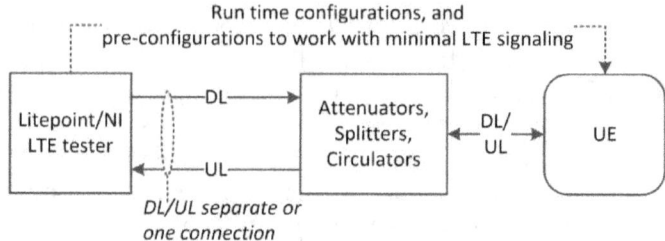

Figure 10.3. UE or modem testing in production.

10.6.4   Extended Lab Testing and Interop

In the laboratory extended test setup shown in figure 10.4, the UE is used similar to the final deployment usage, but with the wireline connections for RF.

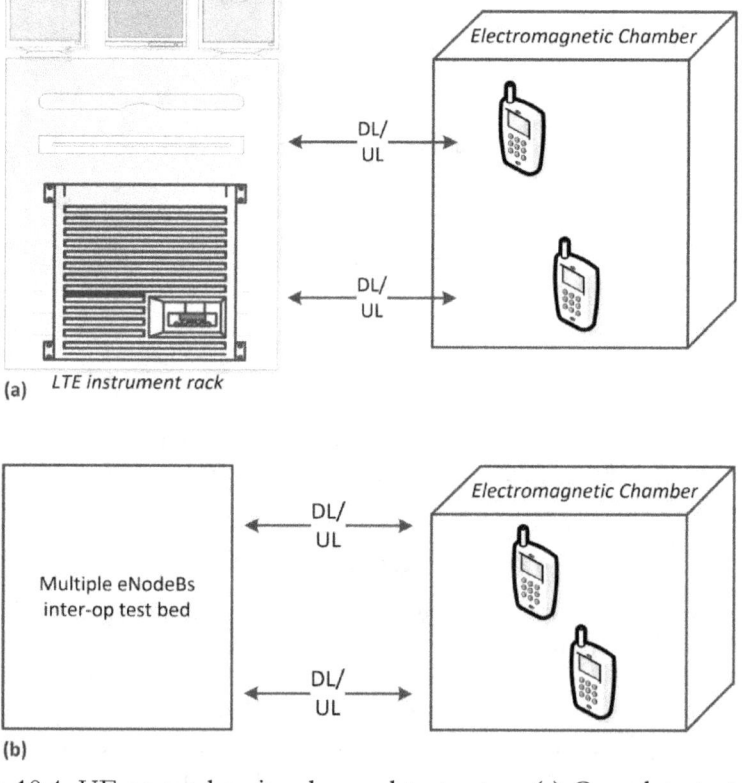

Figure 10.4. UE or modem in advanced test setup. (a) Complete test setup with eNodeB and EPC. (b) Testing with multiple eNodeBs known as interop testing.

The required instrumentation setup is usually in a rack [URL (Anite), URL (Anritsu-7873), Christina Gebner] incorporating various instruments—and many times the instruments from multiple vendors are integrated to achieve the best functionality. The setup also includes channel faders, electromagnetic shields, and temperature variation chambers. As marked in figure 10.4(b), the setup is extended with laboratory environment eNodeBs used with RF cables. Such operation with eNodeBs is referred to as interoperability testing (IOP) and in short form as inter-op [URL (CTIA)].

10.6.5   Certification and Carrier Tests

Several test agencies conduct interoperability testing and carrier specific test suit [URL (Cetecom), URL (CTIA), URL (ODI-Verizon)]. In the test agency setup shown in figure 10.5, the key activities are as follows:

- 3GPP specific tests are conducted.
- Band specific and carrier specific tests are conducted.
- Any industry established new benchmark tests are conducted.

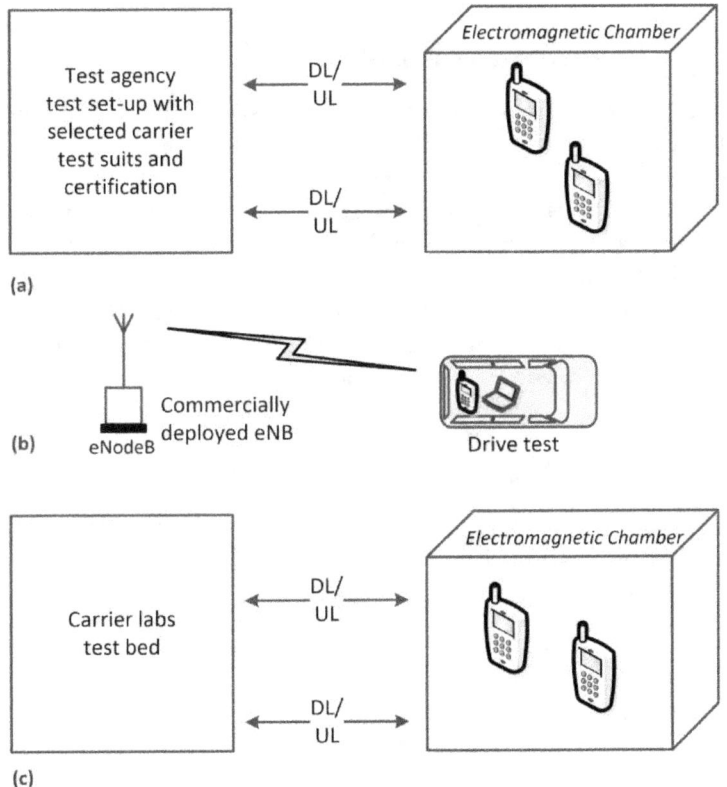

Figure 10.5. UE or modem testing close to deployment scenarios. (a) Test Agency certification. (b) Drive tests. (c) Carrier (service provider) tests.

The test setup shown in figure 10.5(b) is for drive tests in which UE with an extended antenna is taken on a field tests usually referred to as drive tests. The test results are used to further enhance characterization of UE under various conditions. The drive tests need the UE to be compliance with the service network and may need pre-qualification or precertification. With the advancements of fader firmware, such drive test models also got into the fader instruments. The fader instruments [URL (Spirent)] will have utility to select a driving route and various parameters, and the fader loads profile to match the drive test route. This type of setup is advantageous and reduces the test time, expense, and allows repetition and automation.

In test setup shown in figure 10.5(c), UE is tested in a carrier test setup for a few selected tests that are specific to deployment performance, interoperability with other UEs in the network, various time responses, and remote setup configurations. In many occasions carrier may direct to the test agency that replicates carrier requirements and certifies the product [URL (ODI-Verizon)] for deployment.

10.7 Chapter Summary

The book's main focus is the transmission signal quality. The transmission quality cannot be fully concluded with the simple signal generation and analyzer. The setup shown in figure 10.2 followed by setup in figure 10.1 may be a good starting point for arriving at the initial signal quality. The arrived benchmarks can be extended in other setups marked in figures 10.4 and 10.5. The measurements help identify the weakness, and the required quality has to be ensured by design of system architecture, hardware, SoCs, firmware, system integration, engineering the product with SoCs and FEM components including the antennas and enclosures/casing of the product.

11 RADIO LINK BUDGET FOR LTE SYSTEM

LTE systems as per 3GPP specifications are capable of working in the 100 km range [3GPP 36.211, URL (Slideshare-range), URL (pscr)] under favorable conditions. User equipment (UE) physical random access channel (PRACH) preamble format 3 [3GPP 36.211] supports 200 km round-trip (100 km one way) delay. In some selected rural deployments, wide area base station (BS), also known as macro-base-station (BS) or LTE eNodeB, should have capability of working in a higher range approaching 100 km. Refer to [URL (pscr), URL (generaldynamics)] for experiments on achieving the 100 km range. The local BS and home BS work for shorter ranges compared to 100 km. This summary chapter is for providing information on how the LTE 100 km range is established and basic math for the radio link. The calculations are not exact, and the numerical values may deviate from one system design to another, and vary by geography and deployment requirement specifications.

11.1 UE versus eNodeB in Terms of Radio Parameters

The user equipment (UE) is built with lower cost, lower power dissipation, and small form factor. The eNodeB parameters are superior to UE parameters. Link budget is to convey what is the received signal for a given transmitted signal level, under various channel and antenna conditions. As listed in table 11.1 and 11.2, a few key parameter differences for link budget calculation associated with LTE UE and eNodeB are transmit (Tx) power, antenna gain and antenna beam width, penetration losses that also includes antenna efficiency, front-end losses, noise figure of the receiver, downlink (DL) and uplink (UL) frequency difference, and receiver (Rx) sensitivity. With reference to 3GPP 36.942 Table 5.3.3.1, and 3GPP 36.931 specifications, the UE and BS parameters summary is listed in table 11.1.

Table 11.1. UE and BS key parameters considered for range calculation as per 3GPP 36.942

Key parameters	Units	Typical values		Notes
		UE	Macro BS/ eNodeB	
Max Tx power	dBm	23	46	Macro BS power up to 46 dBm, but per each user it is 33 dBm
Per user (UE) max power	dBm	23	33	BS serves multiple UEs, and per UE max is 33 dBm
Antenna gain	dBi	0	15	BS uses directional or beam forming antennas covering the sector
Rx noise figure (NF)	dB	9	5	BS NF is superior
Minimum coupling loss (MCL) urban	dB	70	70	MCL is minimum distance loss measured at antenna connectors, between BS and UE or UE and UE [3GPP 36.942]
MCL rural	dB	80	80	
Most demanded Rx sensitivity at 1.4 MHz	dBm	−106.2	−106.8	UE Band-35/36 lists least power −106.2 dBm, and all other bands are of higher power. For BS it is wide area coverage at 1.4 MHz BW
Rural area antenna height	m	Not applicable	45	From ground reference
Urban area antenna height	m	Not applicable	30	The height 30 m considered as 15 m above average building top or 45 m from ground level
Log normal fading	dB	10	10	Standard deviation is 10 dB
Outdoor wall or penetration loss	dB	10	Not applicable	Penetration loss is applicable at UE side in UL/DL
Beam forming		Not popular	Applicable	Beam forming helps cell edge performance

11.2   REFSENS for Uplink and Downlink

Table 11.2 lists the reference sensitivity (REFSENS) power level for UE and BS. The REFSENS power is denoted as $P_{REFSENS}$. Even though $P_{REFSENS}$ usage may be more correct, commonly REFSENS is used. At long range close to 100 km, the BS may use 1.4 MHz LTE bandwidth (BW) and serves the UE with a less number of resource blocks (RBs), thus creating more power per each LTE resource element (RE) or 15 kHz resolution tone. Referring to table 11.2, Band 13 and Band 14 does not support lower than 5 MHz channel BW. Hence, a less number of RBs may be used to concentrate more power per each tone. The parameters in table 11.2 are compiled from 3GPP 36.101 Table 7.3.1-1, and 3GPP 36.104 Tables 7.2.1-1 to 7.2.1-3. The UE $P_{REFSENS}$ at 1.4 MHz BW is from −106.2 to −101.7 depending on the band. At 20 MHz BW, the UE $P_{REFSENS}$ is from −90 to −94 dBm depending on the band. For base station, the sensitivity is categorized as per wide area, local area and home BS given in 3GPP 36.104.

In chapter 3, noise power is given for multiple LTE-occupied BWs. Relating tables 3.4 and 11.2, the $P_{REFSENS}$ −97 dBm at Band-4 LTE BW 10 MHz and noise in is −104.5 dBm for 9 MHz occupied BW from table 3.4, and typical (whole receiver) UE noise figure (NF) of 5.5 dB; (−97 dBm) − (5.5 dB NF) − (−104.5 dBm for 9 MHz occupied BW from table 3.4) = 2 dB signal power to noise power.

Table 11.2. $P_{REFSENS}$ or REFSENS for UE and BS listed for few selected bands [3GPP 36.101, 36.104]

UE LTE band number	$P_{REFSENS}$ dBm for UE channel BW from 1.4 to 20 MHz						Duplex mode
	1.4 MHz	3 MHz	5 MHz	10 MHz	15 MHz	20 MHz	
4	−104.7	−101.7	−100	−97	−95.2	−94	FDD
7			−98	−95	−93.2	−92	FDD
13			−97	−94			FDD
40			−100	−97	−95.2	−94	TDD
BS category	$P_{REFSENS}$ dBm for BS channel BW from 1.4 to 20 MHz						
Wide Area BS	−106.8	−103.0	−101.5	−101.5	−101.5	−101.5	FDD
Local Area BS	−98.8	−95.0	−93.5	−93.5	−93.5	−93.5	FDD
Home BS	−98.8	−95.0	−93.5	−93.5	−93.5	−93.5	FDD

11.3   Received Power and Path Losses

With reference [URL (antenna-theory)] to equation (11.1), for the DL path BS is the transmitter, P_t is the BS Tx power, and G_t is the BS Tx antenna gain, and G_r is the UE antenna gain. Equation (11.2) indicates loss term with wavelength λ of DL frequency f, c is the velocity of light $(3 \times 10^8$ m/s), and the distance R is between BS and UE. The loss in equation (11.2) is embedded in equation (11.1). The received power at UE is given as follows:

$$P_r = \frac{P_t G_t G_r \lambda^2}{(4\pi R)^2} \text{ or } P_r = P_t G_t G_r (\frac{\lambda}{4\pi R})^2 \text{ or } P_r = \frac{P_t G_t G_r c^2}{(4\pi R f)^2}. \quad (11.1)$$

Direct line of sight path loss (PL).

Path loss in dB $= 10 \log_{10}(\frac{\lambda}{4\pi R})^2.$ (11.2)

As an alternate form of equation (11.2), path loss in dB is given as 20 $\log_{10}(R) + 20 \log_{10}(f) + 20 \log_{10}(4\pi/c)$, where R is in meters, f in hertz, and c is 3×10^8 m/s.

Urban area path loss. As per 3GPP 36.942, the urban area path loss is given by path loss in dB $=$

$40 (1 - 4 \times 10^{-3} \text{ Dhb}) \log_{10}(R) - 18 \log_{10}(\text{Dhb}) + 21 \log_{10}(f) + 80 \text{ dB}$ (11.3)

where R is the base station and UE separation is in kilometers, f is the carrier frequency in MHz, and Dhb is the base-station antenna height in meters—measured from the average rooftop level [3GPP 36.942].

For Dhb of 15 m, the loss at 900 MHz is $120.9 + 37.6 \log_{10}(R)$ dB.

For Dhb of 15 m, the loss at 2,000 MHz is $128.1 + 37.6 \log_{10}(R)$ dB.

Rural area path loss. As per 3GPP 36.942, the rural area path loss as per Hata model is given as:

$69.55 + 26.16\log_{10}(f) - 13.82\log_{10}(\text{Hb}) + [44.9 - 6.55\log_{10}(\text{Hb})] \log_{10}(R) - 4.78(\log_{10}(f))2 + 18.33\log_{10}(f) - 40.94$ (11.4)

where R is the base station–UE separation in kilometers, f is the carrier frequency in MHz, and Hb is the base-station antenna height above ground in meters. For easy math, the following simplified arithmetic is provided for a particular frequency and antenna height that matches to the rural path loss.

- Hb of 45 m, the loss at 900 MHz is given as $95.5 + 34.1 \log_{10}(R)$ dB.
- Hb of 45 m, the loss at 2,000 MHz is given as $102.4 + 34.1 \log_{10}(R)$ dB.
- Hb of 85 m, the loss at 793 MHz is given as $90.75 + 32.26 \log_{10}(R)$ dB.
- Hb of 280 m, the loss at 793 MHz is given as $83.6 + 28.87 \log_{10}(R)$ dB.

The path loss versus distance or range is given figure 11.1 for Band 14 (B14) at 793 MHz. The B14 differs from B13 by 11 MHz (UL 782 MHz in B13, and 793 MHz in B14). In figure 11.1, direct path loss as per Friis transmission equations, urban area path loss for 15 m antenna height from rooftop, rural area path loss at 45 m antenna height are given. The direct path loss is the least and follows inverse square law with distance. The urban path loss is more when compared to rural area path loss.

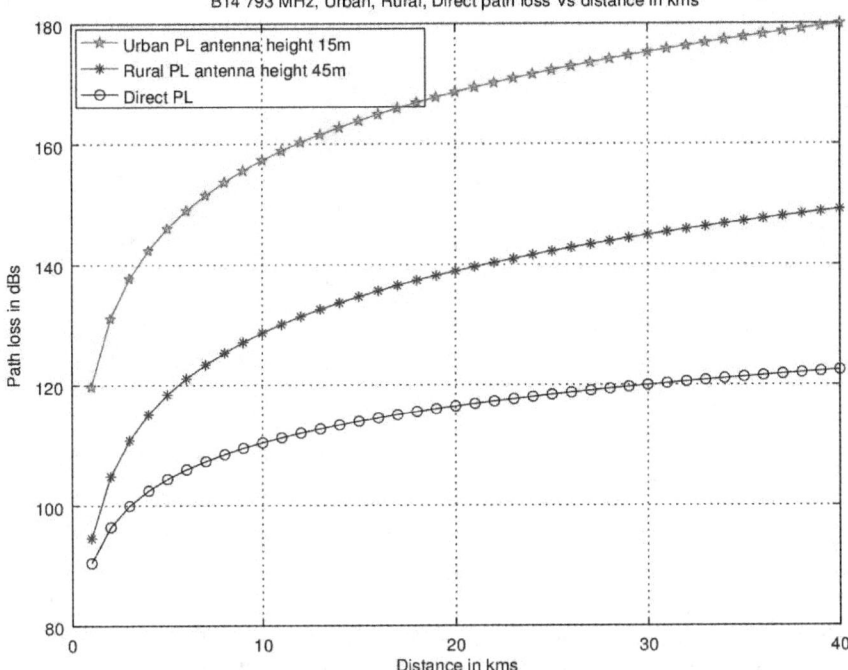

Figure 11.1. Path loss versus range (up to 40 km) at B14 793 MHz carrier frequency—for direct, urban 15 m above rooftop, and rural with 45 m antenna height.

Figure 11.2 is for the 100 km range to create relevance with reported references [URL (pscr), URL (general dynamics)] at Band 14 frequency of 793 MHz. The long distance coverage happens with increased antenna height. Figure 11.2 is shown up to rural 280 m to convey easy interpretation of reduced losses with increased antenna height.

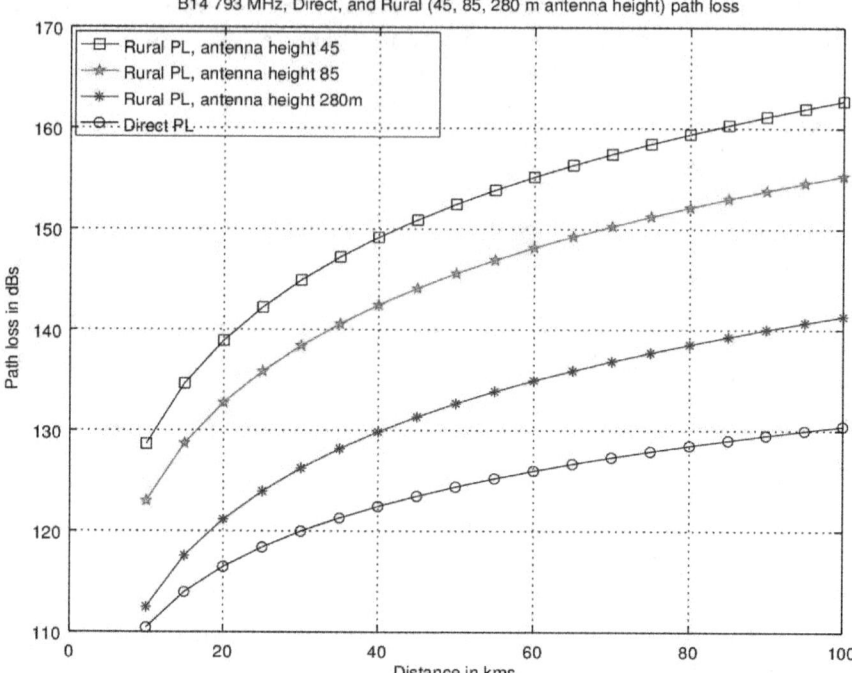

Figure 11.2. Path loss versus range (10–100 km) for direct, and rural with 280, 85, and 45 m antenna height, at B14 793 MHz carrier frequency.

11.4   UL Link Budget

The path loss in equation (11.2) is a direct loss. Table 11.3 lists the UL (UE→BS) transmission at Band 13 (B13) and table 11.4 for Band 4 (B4). The key losses used are direct path loss as per equation (11.2), fading loss of 10 dB, and UE environment or penetration loss of 10 dB. The penetration loss can be widely varying depending on the location of the UE, and in the examples 10 dB is used. As Band 4 is of higher frequency and shorter wavelength, the path loss differs from B13 by 7 dB at 100 km range. The received power at BS receiver is estimated as −119.27 dBm in table 11.4 examples. As per table 11.2, wide area BS requires −106.8 dBm. At this situation, the BS may command UE to use just 1-RB instead of 6 RBs in 1.4 MHz that can give extra $10 \log_{10}(6) = 7.8$ dB benefit, and also the BS may adapt diversity techniques that creates extra gain, and BS may use beam forming techniques to increase antenna gain in the direction of UE.

Table 11.3. UL (UE→BS) link budget for B13 with key loss and gain parameters

Distance in km	1	10	20	50	100
Distance in meters	1,000	10,000	20,000	50,000	100,000
fo MHz in B13 uplink	782	782	782	782	782
Max UE Tx power dBm	23	23	23	23	23
UE antenna gain dB	0	0	0	0	0
UE typical penetration loss dB	10	10	10	10	10
Distance and fo dependent direct path loss in dB	90.36	110.36	116.38	124.34	130.36
Fading loss in dB	10	10	10	10	10
BS Rx antenna gain in dB	15	15	15	15	15
UE Tx to BS Rx input loss dB	95	115	121	129	135
Rx power at eNodeB in dBm	−72.36	−92.36	−98.38	−106.34	−112.36

Table 11.4. UL (UE→BS) link budget for B4 with key loss and gain parameters

Distance in km	1	10	20	50	100
Distance in meters	1,000	10,000	20,000	50,000	100,000
fo MHz in B4 uplink	1,733	1,733	1,733	1,733	1,733
Max UE Tx power dBm	23	23	23	23	23
UE antenna gain dB	0	0	0	0	0
UE typical penetration loss dB	10	10	10	10	10
Distance and fo dependent direct path loss in dB	97.27	117.27	123.29	131.25	137.27
Fading loss in dB	10	10	10	10	10
BS Rx antenna gain in dB	15	15	15	15	15
UE Tx to BS Rx input loss dB	102	122	128	136	142
Rx power at eNodeB in dBm	−79.27	−99.27	−105.29	−113.25	−119.27

Notes: In the previous chapters, higher frequency band examples are shown with Band 7. In this chapter link calculation examples are switched to Band 4, as Band 7 is not listed under 1.4 MHz LTE BW support. To continue examples for 1.4 MHz BW, the description is switched to Band 4. In general Band 7 path loss will be higher compared to Band 4.

11.5 Downlink Link Budget

Table 11.5 lists the DL (BS→UE) transmission at B13 and table 11.6 for B4. The key losses used are path loss as per equation (11.2), fading loss of 10 dB, and UE environment or penetration loss of 10 dB. The wide area BS transmits higher power than UE, and the DL will have advantage for the UE.

Table 11.5. DL (BS→UE) link budget for B13 with key loss and gain parameters

Distance in km	1	10	20	50	100
Distance in meters	1,000	10,000	20,000	50,000	100,000
fo MHz in B13 downlink	751	751	751	751	751
Per UE BS Tx power dBm	33	33	33	33	33
BS antenna gain dB	15	15	15	15	15
UE typical penetration loss dB	10	10	10	10	10
Distance and fo dependent direct path loss in dB	90.01	110.01	116.03	123.99	130.01
Fading loss in dB	10	10	10	10	10
UE Rx antenna gain in dB	0	0	0	0	0
BS Tx to UE Rx input loss dB	95	115	121	129	135
Rx power at UE in dBm	−62.01	−82.01	−88.03	−95.99	−102.01

Table 11.6. DL (BS→UE) link budget for B4 with key loss and gain parameters

Distance in km	1	10	20	50	100
Distance in meters	1,000	10,000	20,000	50,000	100,000
fo MHz in B4 downlink	2,133	2,133	2,133	2,133	2,133
Per UE BS Tx power dBm	33	33	33	33	33
BS antenna gain dB	15	15	15	15	15
UE typical penetration loss dB	10	10	10	10	10
Distance and fo dependent direct path loss in dB	99.08	119.08	125.10	133.06	139.08
Fading loss in dB	10	10	10	10	10
UE Rx antenna gain in dB	0	0	0	0	0
BS Tx to UE Rx input loss dB	104	124	130	138	144
Rx power at UE in dBm	−71.08	−91.08	−97.10	−105.06	−111.08

11.6 Uplink Timing Advance

The electromagnetic waves travel at approximately 3×10^8 m/s, traversing at 300 m/μs. The eNodeB maintains timing and frequency reference. Figure 11.3(a) is for eNodeB sending DL to multiple UEs and farthest UE2 signal undergoes more propagation delay. The 50 km one-way delay is $50,000/(3 \times 10^8) = 166.67$ μs, and the UE1 at 10 km will take 33.33 μs.

As marked in figure 11.3(b), the UE2-DL is delayed with reference to the eNodeB transmission reference. In FDD, the UL is scheduled 4 subframes after the DL subframe. As marked, the UL subframe will go through propagation delay from UE2 to eNodeB. The timing advance (TA) is the round-trip delay that includes eNodeB to UE2 and UE2 to eNodeB with reference to air interface. In figure 11.3(a), UE1 will have shorter timing advance. The eNodeB sends downlink to UE1 and UE2 at the same time reference. The UE1 and UE2 use different time advance values to create time aligned uplink at eNodeB—within the allowed time error/tolerance of 16Ts (0.52 μs).

Key aspects with reference to [3GPP 36.133, 36.321, 36.211] are listed as follows:

- In the initial attach phase, UE establishes DL timing reference and uses the same DL timing reference (aligned at air interface) for sending UL PRACH—matching to air interface.
- While sending PRACH from UE, additional offset N_{TA_offset} (defined as fixed timing advance offset, expressed in units of Ts) is used with value 0 for FDD, and 624Ts for TDD. The 624Ts offset [Ts= 32.55 ns, and 624Ts = 20.3 μs] is marked in TDD for DL to UL transition at eNodeB (eNB).
- Ts is 32.552 ns (numerically 1/30.72 MHz), defined in 3GPP 36.211 as Ts = $1/(15,000 \times 2,048)$ seconds.
- The eNB sends RAR message that provides 11-bit timing advance value T_A to be used by UE.
- As per 3GPP 36.133 section 7.3, the TA command is of 16Ts resolution, and UE has to maintain ±4Ts accuracy while adjusting.
- The T_A value is in 11-bit format and uses from 0 to 1,282; the maximum value 1,282 provides round-trip delay of ($1,282 \times 16$Ts =) 667.7 μs, thus allowing round-trip range of (667.7 μs) × 300 m/μs ≈ 200 km, and one-way range of 100 km.
- The UE calculates required advance N_{TA} in Ts units as $N_{TA} = 16T_A$, and this N_{TA} is usually referred to as N_{TA_old} while updating to new value. N_{TA} [3GPP 36.211] is the timing offset between uplink and downlink radio frames at the UE, expressed in units of Ts.

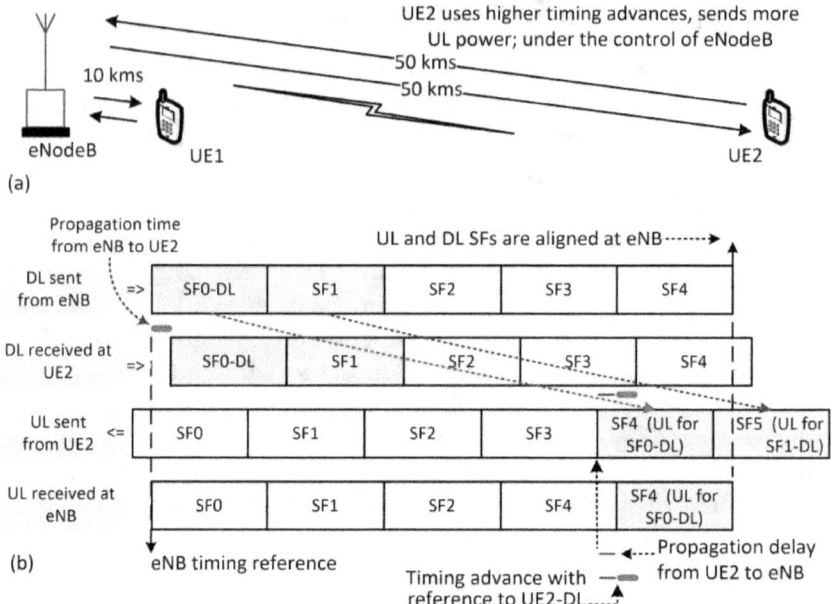

Figure 11.3. Timing advance representation in steady state operation. (a) The eNodeB serving nearby and far-away UEs. (b) Propagation time, timing advance, and alignment at eNodeB shown for steady state timing advance adjusted FDD subframes.

- Once UE is in connected mode, a new T_A value as a 6-bit number is sent from eNB through media access control (MAC) control elements, and UE calculates updated/new value $N_{TA\_new} = N_{TA\_old} + 16 (T_A-31)$ in Ts units. The 6-bit T_A goes from 0 to 63 and (T_A-31) gives from −31 to 32 for positive and negative side timing advance updates.
- Figure 11.3(b) is shown for steady state of timing adjusted subframes. When UE keeps moving away or coming close, a new command may be issued as described in MAC messages [3GPP 36.321]. As per new command, UE shall adjust the UL timing at subframe $n+6$ for a timing advancement command received in subframe n.
- The maximum aggregate adjustment rate given as Tq per 200 ms and Tq is LTE BW dependent. The Tq values for LTE BWs are 1.4 MHz $\Rightarrow$16Ts, 3 MHz $\Rightarrow$8Ts, 5 MHz $\Rightarrow$4Ts, and 10 MHz and 20 MHz $\Rightarrow$2Ts.
- The minimum aggregate adjustment rate shall be 7Ts per second.

Notes: LTE BW 20 MHz uses 30.72 MHz as reference at FFT and IFFT processing. The 30.72 MHz sampling matches with Ts = (1/30.72 MHz). At lower BW 1.4 MHz, the FFT and IFFT are referenced with (30.72/16) MHz sampling that equals to 16Ts. As general interpretation, the TA command with 16Ts works even with 1.4 MHz specific basic sampling of (30.72/16 =) 1.92 MHz. Even though TA command is of 16Ts resolution, the UE should possess capabilities to adjust the timing advance by the Ts interval. The aggregate minimum and maximum are lower than 16Ts, and thus one Ts resolution benefits the adjustments. When LTE BW is changed, the RF FEM, RF, analog processing, and digital processing changes the delay. There could be small timing deviation in RF sections when LTE band is changed. To compensate for these at finer resolution, maintaining timing adjustments at the Ts interval or lower is useful.

11.7   Conclusive Remarks

The link calculations are given here for basic understanding of the end-to-end losses with one low band (Band 13) and one high band (Band 4). The losses will be higher at a higher frequency band. When the link is getting weak, the measurements and feedback mechanisms between BS and UE will adapt to use higher Tx power, lower channel BW, reducing the number of RBs, modifying coding gain (increasing redundancy), using transmit diversity techniques, and adapting to beam forming.

12 IOT SYSTEM AND SIGNAL-QUALITY

The Internet-of-Things (IoT or IOT) cellular/mobile modem is gaining popularity to get wide area and global roaming capability. Many families of IoT devices, such as WiFi, Bluetooth, ZigBee, and Z-wave, work mainly in local area network (LAN) coverage with back-end connectivity to the wide area networks (WAN) through the associated gateway. The cellular-based IoT devices work with cell towers or base stations, thus creating wide area coverage and roaming. In the previous chapters, several topics are covered that include signal analysis and quality improvements for the LTE system/modem. Referring to chapter 1 sections, the basic LTE has six bandwidths (BW) 1.4, 3, 5, 10, 15, and 20 MHz, and the corresponding resource blocks (RBs) are 6, 15, 25, 50, 75, and 100, with each RB is of 12 tones and the modulating tones are spaced at 15 kHz resolution. LTE operates with a minimum BW of 1.4 MHz, and the corresponding occupied BW is 1.08 MHz. During data transmission, LTE can operate even with 1-RB that corresponds to 12 tones and 180 kHz. The narrow-band approach with LTE is to go for 180 kHz occupied BW—usually referred to as 200 kHz channel BW.

The roadmap for LTE, LTE-advanced, and beyond is to create high throughput in hundreds of megabits to gigabits with operational categories from Cat 3 to Cat 10 [URL (LTE-Cat), URL (UE-Cat)], which also uses carrier aggregation [URL (3GPP-CA)]. While the releases progress, the categories and throughput may keep increasing.

The narrow-band (NB) LTE IoT family devices need to operate on a lower BW of 200 kHz to 1.4 MHz. The IoT devices follow discontinuous transmissions and reception with a much shorter ON period and longer OFF period and operate in power-saving modes (PSM). The expected data throughput is in tens of bits to a few kilobits per second.

As narrow-band IoT standards for very low data throughput of operation are further evolving, this chapter is written in a generic way for the narrow-BW LTE system with extended discontinuous reception (eDRX) and transmission with short ON time and long OFF time intervals.

The relevant additional system and signal-quality issues described in this chapter should be applicable commonly for various categories of LTE-based IoT implementations. The general distinguishing characteristics of NB-LTE devices are given as follows:

- Narrow BW operation, and the goal is to keep it in 200 kHz.
- Capability of transmitting high power such as +23 dBm (depending on the narrow-band release), but usually operating in the same LTE Tx power range of −40 to +20 or +23 dBm.
- Higher receive sensitivity (low $P_{REFSENS}$ power) due to low BW, and majorly use lower-order modulations such as quadrature phase shift keying (QPSK).
- Stand-alone battery operation in most applications, and battery should support for a few years of operation.
- IoT devices usually operate in half-duplex mode—between Tx and Rx.
- PSM with eDRX.

Referring to 3GPP 36.802, the following BW and resource element (tone) parameters are applicable for NB-LTE DL and UL:

- Downlink (DL) parameters:
  - 1-RB with 12 tones at 15 kHz resolution.
  - NB-LTE IoT uses 180 kHz RF occupied BW.
  - Carrier spacing is the same, 15 kHz with normal and extended cyclic prefix (CP).
- Uplink (UL) has more tone spacing combinations:
  - 12 tones at 15 kHz resolution.
  - 1 tone at 15 kHz resolution.
  - 48 tones at 3.75 kHz resolution.
  - 1 tone with 3.75 kHz resolution.

The fundamental parameter that benefits NB-LTE link budget (as described in chapter 11) is the narrow BW. The noise power of the ideal receiver is $-174 + 10 \log_{10}(\text{BW})$. With a lower BW of 180 kHz, the noise power is −121.44 dBm. For the same transmitted power (of, say, 20 dBm), the available power for each tone is higher in NB device operation. This translates to the following benefits:

- Ability to communicate to the same distance with very less transmit or receive power.
- Ability to work with higher penetration loss—the IoT device may have higher penetration loss that also includes antenna inefficiencies due to limited size/aperture of the UE antenna.

12.1   LTE-based IoT Categories

LTE IoT relevant categories and summary notes are listed as follows:

- *Cat 1*: The lower throughput category-1 (Cat 1) version was existent from release 8 [URL (LTE-CatM)], and it goes by the name of Cat 1 with lower throughput of DL 10 Mbps and UL 5 Mbps and can operate in the full-duplex mode. The regular modems of Cat 3 and Cat 4 can be easily adapted to work as Cat 1, and Cat 1 may not be suitable for low-power IoT device applications.
- *Cat 0*: The category-0 (Cat 0) is in LTE release 12 with the maximum throughput of 1 Mbps. Cat 0 is also referred to by multiple names, such as Cat-MTC (machine-type communication), LTE-MTC, and LTE-M.
  - Cat 0 uses 1.4 MHz BW and single antenna.
  - Scaling down from Cat 1 to Cat 0 is relatively easier as 1.4 MHz is part of LTE.
  - Cat 0 can be included into main LTE network with the required upgrades in infrastructure.
  - It supports PSM along with eDRX, which is essential for IoT devices [URL (Nokia)] long life.
  - Cat 0 is useful to send media (video, pictures, voice, and audio) clips while performing IoT functions.
- *NB-LTE*: Narrow-band LTE-M is 200 kHz channel BW and a 180 kHz occupied BW LTE system for machine-type communication (MTC) or machine-to-machine (M2M) applications. The narrow-band distinction comes from the use of 200 kHz BW. Refer to [URL (Nokia)] for narrow-band IoT or machine-type wireless/mobile systems and the variants focused around 200 kHz BW. A major part of LTE backhaul may be used with narrow-band operations, but extra base stations may have to be deployed to service NB-IoT devices.

> Notes: While writing this book, the NB LTE-M is a few more months away from the release dates, and generic implementation challenges to achieve signal quality in an NB LTE-M-based products are presented in this chapter.

12.2   IoT Device Hardware Architectural Trade-offs

IoT devices widely vary in their functionality and complexity [David et al, 2012)]. Specific to LTE-based mobile or cellular IoT family, the functional blocks will be in similar lines to chapter 2 of this book, and figures 2.3(c) and (d) provide overview of the architectural split of the low-power LTE

UE devices. IoT device architectures and engineering the product are constrained as follows:

- The cost of IoT modem should be lower than the regular LTE modem.
- Size of the hardware solution has to be much smaller in physical size.
- Power supply source selection, power dissipation, and battery life.
- Requirements of recharging and energy harvesting.
- Ability to handle peak overload of data traffic in emergency/panic situations.
- Number of internal or external sensors and their data size.
- Need of any compressed media (picture, video, voice, and audio) clips data transfer under normal and emergency situations.
- Number of wireless/mobile interfaces, frequency bands to be covered, and RF peak power.
- Silicon technology used/available at the required price point, and low power consumption.
- Antenna size and efficiency for multiple bands of operation.
- Orientation of the IoT device in multiple use cases.
- Stand-alone IoT device loss from antenna connector to the air.
- Penetration loss of the use case environment.
- Electromagnetic absorption, interference, and coexistence from the environment.
- Ability to work in multiple operating orientations (position) with acceptable degradation.
- One-time use or reusable capability.
- Firmware upgradation through remote cloud or one-time programming.
- Generic operating environmental conditions such as extreme thermal (cold or hot) conditions, and many IoT devices have to operate in an industrial-grade environment.
- Hardware operating with much lower leakage currents (even at high temperatures) while sustaining ON/OFF periods.

12.3 Power Dissipation and Peak Power Handling

Relating to the power dissipation of the IoT device, one of the power source options to IoT device is usually a heavy-duty, small-sized battery. The constraints for the battery usage and life span of the battery and IoT device are determined by the following factors:

- Battery capacity.
- Battery self-life loss or leakage.

- Battery equivalent series impedance that drops the voltage at peak loads, and dissipates energy
- Power supply circuit leakage current during nonoperation of the IoT device.
- If IoT device is of interactive type, the frequent overload of panic/emergency situations.
- Power dissipation by IoT device during short operating duration.
- Multiple power rails through power converters and low dropout (LDOs) regulators that create loss of efficiency.
- Energy loss while building up power-good conditions from OFF state.
- Energy loss at the end of ON operation.
- If supercaps (supercapacitors) are used to store energy, the leakage of capacitors contribute to the loss, and supercap equivalent series resistor (ESR) creates voltage drop.
- The supercap charger will have inefficiency.

12.3.1   Peak Tx Power Example

The LTE-based IoT device may operate up to Tx power of +23 dBm (200 mW). To account for RF front-end (RF FEM) losses, additional ~3 dB power is generated in power amplifiers, making RF power generation as +26 dBm (400 mW). With a typical power amplifier (PA) efficiency of 40%, the power drawn from power supply is 1,000 mW. At this high current, the battery or supercap ESR drop will reduce the voltage of the battery at terminals. With power supply efficiency of ~85%, the power drawn from the battery is $1,000/0.85 = 1,176$ mW. As illustrated in chapter 2, the extra digital processing modules will use hundreds of milliwatts during short duration of IoT device operation. Thus overall LTE-based IoT device will demand a big load on commonly used batteries for IoT. To account for such overload, some of the popular implantation techniques buffer extra energy in capacitors such as in supercapacitors. During OFF period, the supercap regains the energy from battery, and during ON period, it supplies higher peak currents. This approach allows use of lower current capacity batteries.

12.4   Power Amplifiers (PA) and Power Supplies

Power amplifier (PA) is part of the LTE UE Tx that delivers required RF power at the antenna port from −40 to +23 dBm. As explained in chapters 2 and 8, the PAs usually provide gain of 6–28 dB in finite steps, and the remaining gain is controlled by the stages preceding the PA. The PA power efficiency and distortions are influenced by the power supply voltage. The PAs are usually provided with two supplies with popular names as V_{BAT}, which is a fixed supply in the range of 3–3.8 V, and a PA VCC is in the range of 0.5–3.6 V. The PA VCC supplies the main source of DC power and has influence on the signal quality.

- In the early products, a fixed VCC power supply of 3.3–3.6 V was commonly used with PA. This is highly inefficient at lower (−40 dBm) Tx power levels. The PA excess power dissipation/loss will be wasted as heat.
- Then comes the average power or average power tracking (APT), where in the power supply voltage changes depending on the scheduled Tx power level. For one orthogonal frequency division multiplexing (OFDM) subframe of 1 ms, the supply voltage is kept constant at a required value. The APT mode of operation is moderately efficient and less complex to implement.
- The better power efficiency option is the envelope-tracking (ET) supply. The envelope-tracking supply follows the high BW OFDM signal envelope. ET-based PAs and power supply need careful effort to minimize distortions, and refer to [URL (NI-PA)] for introductory overview on this topic.

A given LTE system/implementation uses a hybrid combination of many of the above features to cater for various Tx power level changes at LTE symbol resolution. Keeping the DC power loss and heat aside, the fixed voltage and APT-based PAs usually are of good signal quality. In recently years, the ET-based PAs are also engineered to compete in signal quality. There are several factors in signal quality, and first-level two key aspects are error vector magnitude (EVM) and adjacent channel leakage power ratio (ACLR). The EVM is due to various noises and distortions, and ACLR is the spectral distortion.

12.4.1   ACLR

ACLR is the adjacent channel leakage power ratio and is a measure of the amount of power leaking into adjacent channels. ACLR levels appear in adjacent UTRA and E-UTRA (LTE). Refer to [3GPP 36.521-1, Christina Gebner, URL (NI-PA)] for more information on this topic. ACLR is the ratio of the filtered mean power on the assigned channel to the filtered mean power in the adjacent channel. When PA is used in multiple modes with ET, the ACLR becomes the dominant parameter of distortions.

12.5   IoT Device Power Consumption and Battery Life

Estimating IoT device power consumption is a lot more involved process than it is for a regular LTE device. Counting on battery sources, the battery capacity is specified in mAh (milliampere hour), and the voltage rating. For example, consider CR2477 coin cell that has a capacity of ~1,000 mAh ($\Rightarrow$ 1 Ah) at 3 V. In general cell voltage will get lower up to 2 V with time. For energy calculation, the 2.5 V average value may be considered, which gives good estimate. The energy with average voltage of 2.5 V is 1,000 mAh $\times$ 2.5 V = 2.5 $\times$ 1 $\times$ 3,600 (s/h) = 9,000 joules (J).

An LTE system for IoT operates more frequently in receive-mode than transmit. Considering an average power dissipation of 250 mW for a duration of 6 ms will take 6 ms $\times$ 0.25 W = 1.5 mJ energy. For a power conversion efficiency utilization of 85%, the total energy used from battery is 1.5/0.85 = 1.76 mJ. For an average OFF period of 5.12 s, the battery will last for (9,000/1.76$\times$10^{-3}) $\times$ 5.12 s = 26,181,818 s = 303 days. This is just to illustrate narrow-band operation with examples. In general the operating duration will be shorter, and power dissipation at each time will be optimized to provide IoT device operation to last for a few years with a 3 V higher capacity cell.

Processing time and energy. The IoT signal is present for short duration, and the options to consider are to use moderate processing for long duration (within the allowed LTE subframes response time) or use fast processing. The fast processing is more suitable for power sensitive IoT devices as the operations can be completed quickly and the power ON consumption duration can be reduced.

12.6 Supercapacitors

The capacitance of the parallel plate capacitor is given as $\varepsilon A/d$, where ε is dielectric constant (ε is product of dielectric constant of free space ε_0 and relative dielectric constant ε_r), A is area of plates, and d is the separation between plates [URL (hyperphysics)]. For increasing capacitance, distance between plates to be decreased and plates area has to be increased, which is achieved in supercapacitors with special porous material that works as a three-dimensional structure creating equivalent area of several thousand square meters [URL (Cap-xx-FAQ), URL (Ultracaps)] in a few square centimeters of the actual size. The small-sized supercapacitors are available in hundreds of millifarads (mF) to several farads (F) [URL (Cap-xx-products), URL (Ultracaps)]. The batteries used with the IoT product will have limited peak current capability. The supercaps store the energy during the IoT device OFF period and supply peak current/power during the ON period.

12.6.1   Power Supply Overview

The IoT device would need multiple power rails distributed among BBIC, RFIC, PA, RFFEM, external/internal digital/analog interfaces, and sensors. The power management section takes DC voltage and generates multiple rails using the LDOs, buck, and buck-boost converters. Assuming higher current source is available, such as a heavy-duty or rechargeable battery, the power management section derives power directly from the source. When the input is a light (low capacity) current source, such as coin cells, or lower current external power, the power source may charge the supercap, and the supercap handles the peak load. Often there could be low-value capacitors to handle higher-frequency transient currents. The energy harvesters if present will supply very low current that will charge supercap or rechargeable battery. In general energy harvesters may not supply enough current; hence, another primary battery may be used for continuity of the supply. In figure 12.1 multiple options are shown, and all the options will not be present at the same time in one particular product. In general power supplies will come up with some default settings, and processor may take control of the supply through programming interfaces and reconfigures for next time use.

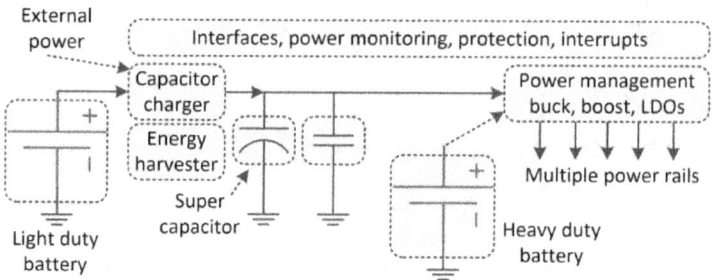

Figure 12.1. Powering the IoT devices.

12.6.2   Capacitor Charging and Energy

The capacitor, C (value in farads), holds the charge Q that has relation with constant current (I_c), voltage (V_c), and time duration (δt). When capacitor acquires charge Q due to flow of current, the following is true:

- The charge in the capacitor is CV_c.
- When constant current (I_c) is flowing for a duration of δt, change in charge is $I_c\,\delta t$.
- When capacitor is charging from 0 to V_c, it takes $\delta t = CV_c/I_c$ to reach V_c voltage.
- The energy stored in capacitor is given as $\frac{1}{2}\,CV_c^2 = \frac{1}{2}\,QV_c$.

The following example is formed with 1,000 mF capacitance and 10 mA charge current. The capacitor charges to 2.5 V, discharges to 2.2 V, and recharges to 2.5 V:

- A supercap of 1,000 mF charging from 0 to V_2 = 2.5 V with 10 mA constant current will take time for charging $\delta t = CV_2/I_c$ = 1,000 mF × 2.5 V/10 mA = $1,000 \times 10^{-3} \times 2.5/(10 \times 10^{-3})$ = 250 s. The energy in the capacitor at the end of charging is $\frac{1}{2} CV_2^2 = \frac{1}{2}$ 1,000 mF × 2.5^2 = 3.125 J.
- The energy remaining in the capacitor when it discharges to V_1 = 2.2 V is $\frac{1}{2} \times$ 1,000 mF × 2.2^2 = 2.42 J, and the energy change 0.705 J can be expressed as $\frac{1}{2} C (V_2^2 - V_1^2)$.
- Considering an example of 1 W power dissipation for an IoT device, the device can sustain for approximately 0.705 s (without counting ESR and power conversion losses), while voltage is dropping from 2.5 V to 2.2 V on supercaps. For 250 mW power dissipation example, the charge can support approximately 4 × 0.705 = 2.82 s.
- The time to recharge the capacitor from V_1 = 2.2 to V_2 = 2.5 V is $\delta t = C (V_2 - V_1)/I_c = 1,000 \times 10^{-3} \times (2.5 - 2.2)/(10 \times 10^{-3})$ = 30 s with a current of 10 mA. Without accounting any inefficiencies, 30 s charge supported 2.82 s of operation (at 250 mW) in the numerical examples.

12.6.3   Energy Lost in Charging Resistor

When the capacitor is charged from a battery, the equivalent series-charging resistor (active or passive) between the battery and the capacitor is R_c. At initial charge (first time power ON), energy transferred to capacitor is $\frac{1}{2} CV_c^2$, and the loss in resistor is $\frac{1}{2} CV_c^2$, a loss of 50% in the resistor. This will not change with the value of the resistor; refer to [URL (hyperphysics), URL (Cap-xx-Power2)] for analysis/clarifications on capacitor charging. Once capacity is partly discharged, the capacitor recharging loss from level V_1 to V_2 depends on the voltage difference $(V_2 - V_1)$, and by maintaining the difference as small voltage, the losses can be reduced. In the previous section, the charging current is assumed as a constant for easy arithmetic. The actual charging current may be exponentially decreasing on the basis of the voltage gap between V_2 and V_1.

> Notes: The use of direct energy from battery or any other direct source will be more efficient. The supercap charger or series-charging resistor will have loss. The equivalent series resistance of a supercap and capacitance leakage create additional loss. The main benefit of a supercap is that it can supply higher peak currents and can store energy sufficient for a few seconds of continuous operation, which may not be sustainable with small-capacity batteries.

12.6.4   Supercap Ceff and ESR

The equivalent capacitance of a supercapacitor (Ceff) is dependent on pulse width (time-duration) of use. Wireless systems operate from 100 μs to a few milliseconds. The Ceff is the percentage of DC capacitance at a given pulse width. The Ceff value improves with pulse width, which can be noted from [URL (Cap-xx-ceff)] as 1 ms 14%, 10 ms 41%, and 100 ms 80%. Considering an LTE Tx subframe of 1 ms, the capacitance of 1,000 mF will appear as 14% of 1,000 mF = 140 mF. The charge in the capacitor is not lost, and it is completely present in the capacitor. The Ceff is the indication of time-frequency response to short pulsed currents. While doing system analysis, the Ceff has to be taken into account. When 1 ms pulse current is used, the ESR will also drop the terminal voltage, and Ceff will make the capacitor to behave like a small-value (140 mF out of 1,000 mF for 1 ms duration) capacitor, and Ceff improves with longer duration operation.

12.6.5   Supercap Tolerance, ESR, and Temperature

Supercaps are less precise in the capacitance value, and they have higher tolerance of ±20% to ±30%. The ESR of a supercapacitor is temperature dependent. The industrial-grade IoT devices may operate from −40 °C to +70/85 °C. Referring to capacitance data sheet [URL (Cap-xx-ceff)] as an example, the ESR at −40 °C is approximately 2.2 times the ESR at room temperature, and ESR at +70 °C is approximately 0.8 times the ESR at room temperature. The RF PA circuits (and in general semiconductor chips) usually consume more current at a higher temperature. The low ESR of a supercap at a higher temperature helps this condition.

12.6.6   Single versus Dual-Cell Supercaps

The single-cell supercaps for IoT applications operate close to 2.7 V (as per current technology). The dual-cell is a series connection of two single cells. The series connection of two 1,000 mF capacitors creates one 500 mF capacitance with 2×2.7 V voltage capacity. The ½ CV_c^2 of a dual cell [½ (500 mF) (2×2.7)^2] is twice that of a single cell [½ (1,000 mF) (2.7)^2]. The dual-cell capacitors are with three terminals and need special care in balancing the charge; otherwise they will get damaged. The simple balancing is with resistor load as marked in [URL (Cap-xx-products)], and it is lossy solution for IoT devices. Chips [URL (LTC-Balancer)] are available with built-in charging and balancing to avoid loss in resistive balance.

12.6.7   Energy Harvesting and Leakage

In some IoT product deployments energy harvesting is possible, such as from solar, vibration, heat, and electromagnetic induction. The harvesting provides tiny energy over long duration that has to be accumulated into chargeable cell or small-capacity supercaps. When energy harvesting is used with IoT devices, in many occasions a primary storage cell may be required to continue the operations when harvested energy is not sufficient or to handle emergency situations. Energy harvesting chargers may work even with hundreds of nanoamperes current. In this situation the whole power supply system has to be properly analyzed for leakage of the batteries, supercaps, and the nonlinearity to accept small currents charging. Supercaps leak the charge, and the leakage varies from a fraction of microampere to tens of microamperes, which is significant in energy harvesting applications.

12.7   Narrow-Band LTE Signal Levels

The signal-quality aspects and analysis in the previous chapter are also relevant to the narrow-band LTE IoT device with some key deviations listed below:

- IoT device is half duplex that avoids the interference from Tx to Rx through conducted coupling and radiated coupling; this is a benefit of the IoT relative to the FDD-based LTE operation.
- Referring to chapter 3, the peak-to-average-power ratio (PAPR) or the crest factor of narrow-band operation will be lower than the regular LTE BW signals. The lower PAPR eases the operation of power amplifiers peak power handling, and average power tracking mode power supply can be made to provide lower voltage. The analog to digital and digital to analog converters (ADCs/DACs) will also be of a lower number of bits compared to the regular LTE BW of operation.
- The receivers have to operate with a lower signal level ($P_{REFSENS}$) due to lower BW.

12.8   Narrow-Band—Frequency and Clock Sources

Refer to chapter 4 for better continuity of this section. The UE acquires the frequency reference through primary synchronization signal (PSS), and reference signals (pilot tones). The accuracy of such clock reference improves on the basis of the duration of the available LTE signal over several slots, subframes, or frames. Narrow-band operation has more difficulties compared to normal LTE operation as described below:

- The analysis window is limited to either a few symbols in a subframe or to a few subframes for reducing ON duration power dissipation, and this limits the accuracy of the recovered frequency reference.
- The device may not use temperature-controlled XO (TCXO) due to cost constraints, and instead lower ppm crystal oscillator (XO) may be

used.

- The crystal oscillator will have quick frequency drifts when operating in varying temperature environment.
- Once frequency reference is established on the basis of the received signal and crystal oscillator, the same reference or derived frequency errors have to be used to create Tx LO frequency. The LTE Tx has to be maintained to within ±0.1 ppm. This will impose multiple difficulties as listed:
  - Due to half-duplex operation, the Tx would operate in a different subframe than Rx reference, and this delay may create frequency drift.
  - The Rx and Tx will operate for short duration, and OFF period is 1.28 to tens of seconds. During this OFF period, the clock reference may drift; hence, references need to be reestimated before attempting the transmissions, making it more stringent constraints than regular LTE operation.
  - Due to Doppler change in the OFF period, the previously established frequency reference may not be suitable for current subframe processing.

12.8.1   Synthesizers

The synthesizer schemes presented in sections 4.4 and 4.5 are still applicable to narrow-band operation. Due to half-duplexing operation, the Tx and Rx may share the same synthesizers. One of the difficulties with synthesizers will be with waking up. On waking up, the synthesizer takes several tens of microseconds to settle, and to reach to less than ±0.1 ppm the power is lost in hardware modules. The crystal oscillators or TCXOs if shut down may take 2–10 ms to reach stable state of operation, and then synthesizer will start settling.

12.9   Narrow-Band—Slot and Frame Alignment

In LTE, first-level slot and frame alignments are established using secondary synchronization signal (SSS) and PSS signals. The timing alignment is further enhanced using the channel impulse responses as part of time and frequency-domain processing. The key aspects of narrow-band operations are listed below:

- The established timing alignment has to be maintained during the OFF period of several seconds.
- If the precision timer with LTE reference clock synthesizers are active that will be a lot of power consumption during the OFF period.
- If real-time clock (RTC) is used, that will be less accurate (than LTE

reference) and the established time alignment will deviate.

- The IoT typically wakes up at the specified paging intervals to analyze the reception. If the reception is not for that particular IoT device, it goes to various modes of sleep. The accuracy and resolution of wake-up timer will decide on how early the device should wake up to account for wake-up timer errors and resolution.
- LTE system waking up and reestablishing the time alignment will be loss of computation and power dissipation.
- If LTE system changes relative motion that can deviate in frequency due to Doppler, 0.1 ppm allowed error for uplink will not be accurate enough.

12.10 Narrow-Band—Gain Controls

Chapter 9 has the basic approach for the LTE signal level estimation and gain control. The IoT device wakes up, and the initial gain control may be based on the previous gain. The new subframe power immediately after waking up may be of different characteristics in level. To counter this, time-domain signal strength estimation has to be reestablished. Even though the PAPR will be low for narrow-band LTE signals, the suggested approach would be to operate with a good dynamic range up to ADC output that allows headroom to minimize signal clipping or getting submerged in the noise floor.

12.11 IoT Device Size and Antenna Size

IoT devices usually of a few square inches surface area. Considering LTE Band 13 at 751 MHz, the wavelength λ in free space $= c/f$, where c is light velocity $= 3{\times}10^8$ m/s, and f is the frequency in hertz. For B13, $\lambda = 3{\times}10^8/(751 \text{ MHz}) \approx 0.4$ m or 40 cm or 15.72 inches. The antenna has to be comparable to $1/4$th of the wavelength, which is ~4 inches for B13. The 4 inch length is not possible in many IoT devices. The wavelength in nonfree space is shorter and influenced by the dielectric constant. The wavelength with ceramic or ceramic chip antennas is lower by several factors on the basis of relative dielectric constant ε_r [URL (ursi)]. The modified λ denoted here as λ_m is given as $\lambda/\text{sqrt}(\varepsilon_r)$. Refer to [URL (taoglas-internal), URL (JOHANSON)] for more details on these antennas and their characteristics. The small antennas do need certain clearances and ground planes while mounting on printed circuit board (PCB) to obtain the stated efficiency.

12.11.1 Active Antenna and Tuning

User equipment (UE) antennas are typically classified as external and internal to the UE modem. The choice depends on multiple factors, and the main factors being the space, cost, and use case scenarios. The tiny IoT and smart phones are built with internal antennas; the tablets can be with internal or external antenna, and automobile LTE systems may use external antenna. There are many types of external and internal antenna. Considering example antennas in [URL (taoglas-external), URL (taoglas-internal)], eight different categories of external antennas and five different categories of internal antennas are listed for LTE, and each manufacturer [URL (ethertronics2)] has its own derivatives for a particular quality benefit and market needs [URL (lstelcom)]. The internal antennas are usually constrained by available space, could be lossy and less efficient, and may be limited in control on the antenna pattern. Even though UE antenna are specified as omnidirectional, they will have partial direction characteristics, and depending on the way LTE modem is physically mounted and held, it will influence the reception/transmission.

As per 3GPP quality metrics, the signal quality for UE is measured at the antenna ports. The antenna influence will be external to chapter 10, tables 10.1 and 10.2. The antenna losses are accounted as part of coupling loss that includes Tx antenna characteristics, penetration loss, path loss with various fading situations, and receiver (Rx) antenna characteristics.

Referring to chapter 2 on RF FEM receive path, the signal starting from the antenna port under goes loss in the FEM and further gets degraded in the low-noise amplifier (LNA) by its equivalent noise figure (NF). The losses and NF combined are of the order of 5–6 dB in a typical LTE UE modem. Added to this a less efficient antenna can create signal-quality degradation.

External active antenna [URL (ethertronics1), URL (ethertronics2), URL (Commscope)] are commonly used in base stations. External active antenna with UEs may use separate antenna for Tx, and Rx. The Rx active antenna is used with superior quality LNA with noise figure of 1.2–1.5 dB that becomes part of the active antenna. The same antenna RF cable or printed circuit board (PCB) trace from FEM to antenna is used to power the antenna-LNA. Overall the FEM losses and main LNA (part of RFIC in chapter 2 examples) noise figure are reduced with active antenna. The antenna-LNA noise figure will be the main dominant contributor, which is much lower compared to RF integrated circuit (RFIC)-LNA and FEM losses.

The second aspect of improvements is through antenna tuning. The tuning is to optimize the performance in multiple aspects, and to name a few first-order benefits, it adapts depending on how the user is holding the UE or how it is mounted on a particular product and changes in matching/coupling with band change. As such the antenna is not intelligent to change its matching circuit. The matching has to be adaptively sensed, and adjustments have to be made. The simplest form of such adjustments may be just switching a different value of tuning inductors and capacitors. As a consolidated summary, to improve on the quality various options to consider are as follows:

- Short or small (relative to wavelength) sized antennas tend to be less efficient.
- RF FEM losses influence can be minimized with active antenna.
- The distance from antenna and RF FEM PCB traces or cable length can increase the loss. This will not be a serious consideration with active antenna.
- Antenna tuning improves the coupling or matching, thus transferring maximum power.
- If the system/product has multiple wireless modules, their coexistence for various conducted and radiated coupling has to be well understood, and interference has to be minimized. Antenna location and orientation will have influence on the coexistence.

12.12 IoT Device Testing

The LTE testing approach is presented in chapter 10. The short-duration ON period of an IoT device has special needs, which makes it more difficult to test compared to a regular LTE UE device. The regular UE device interacts with the base station and infrastructure that can be emulated through various instruments. The LTE UE device is majorly power ON and active throughout the tests controlled by the base station. The IoT device goes ON/OFF as decided by both the base station and the UE. The IoT device, while interacting with various sensors, may be in need of connectivity, and it may start interacting with the base station, overriding the previously set configurations. The instrumentation has to account for such situations. As there are many versions and deviations in IoT overall product requirements, there will be certain amount of proprietary nature of the instrumentation. The IoT devices go to sleep state—meaning the device will be inactive for hundreds of milliseconds to a few seconds before receiving the next useful signal. The device can also go to deep sleep—tens of seconds to a few minutes of nonoperation. All the sleep states may not be standard compliant and vary widely from one implementation to another. As additional generic requirements, the transient responses have to be validated for the following:

- The power loss while device is entering sleep and deep-sleep states.
- Power loss while the device is waking up.
- The power supplies, frequency, and timing references have to be

validated for restarting within acceptable limits before the actual ON period.

- The housekeeping while going to sleep, preservation of the required configurations, and reapplying of these at the right time after waking up.
- Validity of precalibrated configurations during waking up.

12.12.1 IoT Device Coexistence

A major portion of IoT devices stay stand-alone and communicate with the local gateway through Bluetooth, WiFi, LTE, and many other proprietary wireless/mobile communication methods. IoT devices have short-duration ON/OFF and may not be monitoring and analyzing the activity of other radiations. If the IoT is a stand-alone device, it should have enough filters and band spectrum avoidance to protect itself from other interferers and continue for acceptable operation. In a simpler implementation, the IoT device may keep good surface acoustic wave (SAW) family filters that will allow required band signals.

12.12.2 IoT Device Environmental Conditions

IoT device environmental conditions vary widely depending on the end application. A particular class of devices may operate in just ambient temperature, and another class may be serving the industrial environment. The IoT, when it is supposed to serve the severe conditions such as fire or floods, will have several other factors. In general, some class of IoT devices may need to work in severe weather conditions and outdoor operations.

12.13 Signal-Quality Miscellaneous Connections

LTE tests as per 3GPP 36.521-1, sections 6 and 7, are listed in tables 10.1 and 10.2. These are more close to the transmission characteristics and the performance under various constraints to match the usage environment. Here is a conclusive summary:

- The parameters and their variants with amplitude, phase, time, frequency, and spatial characteristics (propagation losses, and fading) influence the signal quality.
- The hardware and firmware controlling the parameters and their adaptability can help change the dynamics of signal quality.
- Once hardware is involved with RF and analog, the process variations, voltage, temperature (PVT), and mechanical vibrations and shocks also influence the quality.
- The antenna orientation, patterns, efficiency (losses), directivity, incorporation of antenna and RF FEM tuning, active antenna

provision, antenna adaptive tuning, electromagnetic environment UE is operating can also moderate the signal quality.

- When multiple wireless devices are coexisting in the proximity, the wireless devices/protocols that share the time slots can significantly influence the signal quality.
- The transmit ON/OFF masks in 3GPP are a bit more liberal in specifications, and practical systems have to be built with better specifications, which makes the difference.
- The nonlinearity and intermodulation (IM) specific to signal levels and various noises are also important.
- When the LTE system is operating in multiple radio access technology (RAT), power dissipation, supply voltage, and temperature are important aspects that can influence the system.

Bibliography

References by Standards

3GPP TS 24.301 version 9.11.0 Release 9. Non-access-stratum (NAS) protocol for evolved packet system.
https://www.etsi.org/deliver/etsi_ts/124300_124399/124301/09.11.00_60/ts_124301v091100p.pdf

3GPP TS 24.302 version 9.7.0 Release9. Access to the 3GPP evolved packet core (EPC) via non-3GPP Access Networks.
https://www.etsi.org/deliver/etsi_ts/124300_124399/124302/09.07.00_60/ts_124302v090700p.pdf

3GPP TS 36.101 version V9.22.0 Release 9. User Equipment (UE) radio transmission and reception.
http://www.etsi.org/deliver/etsi_ts/136100_136199/136101/09.22.00_60/ts_136101v092200p.pdf

3GPP TS 36.104 version 9.13.0 Release 9. Base Station (BS) radio transmission and reception.
http://www.etsi.org/deliver/etsi_ts/136100_136199/136104/09.13.00_60/ts_136104v091300p.pdf

3GPP TS 36.133 version 9.22.0 Release 9. Requirements for support of radio resource management.
http://www.etsi.org/deliver/etsi_ts/136100_136199/136133/09.22.00_60/ts_136133v092200p.pdf

3GPP TS 36.201 version 9.1.0 Release 9. General description.
http://www.etsi.org/deliver/etsi_ts/136200_136299/136201/09.01.00_60/

3GPP TS 36.211 version 9.1.0 Release 9. Physical channels and modulation.
http://www.etsi.org/deliver/etsi_ts/136200_136299/136211/09.01.00_60/ts_136211v090100p.pdf

3GPP TS 36.212 version 9.4.0 Release 9. Multiplexing and channel coding.
http://www.etsi.org/deliver/etsi_ts/136200_136299/136212/09.04.00_60/ts_136212v090400p.pdf

3GPP TS 36.213 version 9.3.0 Release 9. Physical layer procedures.
http://www.etsi.org/deliver/etsi_ts/136200_136299/136213/09.03.00_60/ts_136213v090300p.pdf

3GPP TS 36.214 version 9.2.0 Release 9. Physical layer measurements.
http://www.etsi.org/deliver/etsi_ts/136200_136299/136214/09.02.00_60/ts_136214v090200p.pdf

3GPP TS 36.321 version 9.6.0 Release 9. Medium access control (MAC) protocol specification.
http://www.etsi.org/deliver/etsi_ts/136300_136399/136321/09.06.00_60/ts_

136321v090600p.pdf

3GPP TS 36.322 version 9.3.0 Release 9. Radio link control (RLC) protocol specification.
http://www.etsi.org/deliver/etsi_ts/136300_136399/136322/09.03.00_60/ts_136322v090300p.pdf

3GPP TS 36.323 version 9.0.0 Release 9. Packet data convergence protocol (PDCP) specification.
http://www.etsi.org/deliver/etsi_ts/136300_136399/136323/09.00.00_60/ts_136323v090000p.pdf

3GPP TS 36.331 version 9.18.0 Release 9. Evolved universal terrestrial radio access (E-UTRAN); radio resource control (RRC) protocol specification.
http://www.etsi.org/deliver/etsi_ts/136300_136399/136331/09.18.00_60/ts_136331v091800p.pdf

3GPP TS 36.401 version 9.2.1 Release 9. Evolved universal terrestrial radio access network (E-UTRAN), architecture description.
http://www.etsi.org/deliver/etsi_ts/136400_136499/136401/09.02.01_60/ts_136401v090201p.pdf

3GPP TS 36.521-1 version 9.7.0 Release 9. User Equipment (UE) conformance specification; radio transmission and reception; Part 1: Conformance testing.
http://www.etsi.org/deliver/etsi_ts/136500_136599/13652101/09.08.00_60/

3GPP TR 36.931 version 9.0.0 Release 9. Radio frequency (RF) requirements for LTE pico NodeB.
https://www.etsi.org/deliver/etsi_tr/136900_136999/136931/09.00.00_60/tr_136931v090000p.pdf

3GPP TR 36.942 version 9.3.0 Release 9. Radio frequency (RF) system scenarios.
https://www.etsi.org/deliver/etsi_tr/136900_136999/136942/09.03.00_60/tr_136942v090300p.pdf

ITU-R SM.329-10 (02/2003). Unwanted emissions in the spurious domain.
https://www.itu.int/dms_pubrec/itu-r/rec/sm/R-REC-SM.329-10-200302-S!!PDF-E.pdf

References by Author Index

Afsaneh Nassery, Srinath Byregowda, and Sule Ozev, (2012), Built-in-self test of transmitter I/Q mismatch using self-mixing envelope detector, 2012 IEEE 30th VLSI Test Symposium (VTS).

Award Solutions, Mastering the LTE air interface, 4G LTE curriculum, LTE University, Award Solutions.

Chris Johnson (2012), Long Term Evolution in Bullets, second edition.

Christina Gebner, Long Term Evolution, A concise introduction to LTE and its measurement requirements. Rohde & Schwarz.

David Boswarthick, Omar Elloumi, Oliver Hersent, (2012), M2M

Communications A Systems Approach, John Wiley & Sons.

Edmund Coersmeier, Ernst Zielinski, Frequency selective IQ phase and IQ amplitude imbalance adjustments for OFDM direct conversion transmitter. Accessed 2016 June 4 at: http://www.its.bldrdoc.gov/isart/art03/slides03/coe_e/coe_paper.pdf

Erik Dahlman, et al, (2011). LTE/LTE-advanced for mobile broadband, Academic Press, UK

Moray Rumney (2013). LTE and evolution to 4G wireless, design and measurement challenges, second edition, Wiley, USA. 2013.

Robins W.P (1984). Phase noise in signal sources, IEE Telecommunications series 9.

Sivannarayana, N. and Rao, K.V. (1998). I-Q imbalance correction in time and frequency domains with application to pulse Doppler radar. Sadhana, 23, 1998, pp. 93–102.

Stefania Sesia, Issam Toufik, Matthew Baker (2011), LTE - The UMTS Long Term Evolution: From theory to practice, second edition, Wiley.

References by URLs

URL (3GPP-CA). Carrier aggregation explained. Accessed 2016 June 4 at: http://www.3gpp.org/technologies/keywords-acronyms/101-carrier-aggregation-explained

URL (3GPP-RAN). Radio access network. Accessed 2016 June 4 at: http://www.3gpp.org/specifications-groups/ran-plenary

URL (AD9361). AD9361, RF agile transceiver. Accessed 2016 June 4 at: http://www.analog.com/media/en/technical-documentation/data-sheets/AD9361.pdf

URL (ADI1). Converting oscillator phase noise to time jitter. Accessed 2016 June 4 at: http://www.analog.com/media/en/training-seminars/tutorials/MT-008.pdf

URL (ADI2). Aperture time, aperture jitter, aperture delay Time—removing the confusion. Accessed 2016 June 4 at: http://www.analog.com/media/en/training-seminars/tutorials/MT-007.pdf

URL (ADI-AD9361). RF agile transceiver AD9362. Accessed 2016 June 4 at: http://www.analog.com/media/en/technical-documentation/data-sheets/AD9361.pdf

URL (Aeroflex-7100). 7100 digital radio test set. Accessed 2016 June 4 at: http://ats.aeroflex.com/mobile-device-test/mobile-device-test-products/7100-digital-radio-test-set

URL (Agilent-CCDF). Characterizing digitally modulated signals with CCDF curves. Accessed 2016 June 4 at: http://www.emce.tuwien.ac.at/hfadmin/354059/download/RF4%20Characterizing%20Digitally%20Modulated%20Signals%20with%20CCDF%20Curves.pdf

URL (ALU1). The LTE network architecture. Accessed 2016 June 4 at: http://www.cse.unt.edu/~rdantu/FALL_2013_WIRELESS_NETWORKS/LTE_Alcatel_White_Paper.pdf

URL (Anite). LTE performance testing. Accessed 2016 June 4 at: http://www.anite.com/businesses/handset-testing/lte-performance-testing

URL (Anritsu-7873). LTE RF test conformance system. Accessed 2016 June 4 at: http://www.anritsu.com/en-US/test-measurement/products/ME7873L

URL (Anritsu-8820C). MT8820C Radio Communication Analyzer. Accessed 2016 June 4 at: http://dl.cdn-anritsu.com/en-au/test-measurement/files/Product-Introductions/Product-Introduction/MT8820C_EL1900.pdf

URL (Anritsu-8821C). Radio communication analyzer. Accessed 2016 June 4 at: https://www.anritsu.com/en-US/test-measurement/products/mt8821c

URL (Anritsu-LTE). Long Term Evolution – LTE/LTE-Advanced. Accessed 2016 June 4 at: https://www.anritsu.com/en-US/test-measurement/technologies/lte

URL (Anritsu-measurements). LTE measurements radio communication analyzer MT8820C/8821C. Accessed 2016 June 4 at: https://dl.cdn-anritsu.com/en-en/test-measurement/files/Application-Notes/Application-Note/mt8820c-21c-ef6300.pdf

URL (antenna-theory). Antenna basics. Accessed 2016 June 4 at: http://www.antenna-theory.com/basics/friis.php

URL (Azimuth). Environment emulator. Accessed 2016 June 4 at: http://www.azimuthsystems.com/products/ace-rnx/

URL (Berkeley). Receiver architectures. Accessed 2016 June 4 at: http://rfic.eecs.berkeley.edu/~niknejad/ee242/pdf/eecs242_lect3_rxarch.pdf

URL (BiST). Built-in-self test of transmitter I/Q mismatch and nonlinearities using self-mixing envelope detector, MS thesis. Accessed 2016 June 4 at: http://repository.asu.edu/attachments/94004/content/tmp/package-D7owfS/Byregowda_asu_0010N_12067.pdf

URL (Cap-xx-ceff). GW103 / GW203 supercapacitor. Accessed 2016 June 4 at: http://www.cap-xx.com/wp-content/uploads/2015/10/CAP-XX-GW103-GW203-Datasheet-v4-1.pdf

URL (Cap-xx-FAQ). Supercapacitor FAQ. Accessed 2016 June 4 at: http://www.cap-xx.com/products/faqs/

URL (Cap-xx-Power2). Using a supercapacitor to power wireless nodes from a 3V button battery. Accessed 2016 June 4 at: http://www.cap-xx.com/wp-content/uploads/2015/04/Using-a-Supercapacitor-to-Power2.pdf

URL (Cap-xx-products). CAP-XX supercapacitors product guide. Accessed 2016 June 4 at: http://www.cap-xx.com/wp-content/uploads/2015/10/CAP-XX-Product-Guide-2015-v3.0.pdf

URL (Cetecom). Mobile communications. Accessed 2016 June 4 at:
http://www.cetecom.com/fileadmin/docs/PDF/Mobile_Communications_.p
df

URL (Charon). Tutorials on Digital Communication Engineering.
Accessed 2016 June 4 at:
http://complextoreal.com/tutorials/#.V4HUQLgrLIU

URL (Commscope). Active antennas: The next step in radio and antenna
evolution. Accessed 2016 June 4 at:
http://www.andrew.com/docs/active_antenna_system_white_paper_wp-
105435.pdf

URL (CQI). LTE quick reference CQI. Accessed 2016 June 4 at:
http://www.sharetechnote.com/html/Handbook_LTE_CQI.html

URL (Crystek). Sources of phase noise and jitter in oscillators. Accessed
2016 June 4 at:
http://www.crystek.com/documents/appnotes/SourcesOfPhaseNoiseAndJitte
rInOscillators.pdf

URL (Crystek-TCXO). Understanding TCXOs. Accessed 2016 June 4 at:
http://www.crystek.com/documents/appnotes/Understanding_TCXOs.pdf

URL (CTIA). CTIA certification, test plan for LTE interoperability.
Accessed 2016 June 4 at: http://www.ctia.org/docs/default-source/default-
document-library/ctia-test-plan-for-lte-interoperability.pdf?sfvrsn=0

URL (Delay1). Phase delay Vs. Group delay. Accessed 2016 June 4 at:
http://www.markimicrowave.com/blog/2013/11/phase-delay-vs-group-delay/

URL (dline-CFO). Carrier frequency synchronization in OFDM-downlink
LTE systems. Accessed 2016 June 4 at:
http://www.dline.info/jes/fulltext/v4n3/2.pdf

URL (Embedded). Digital signal processing tricks–high speed vector
magnitude approximation. Accessed 2016 June 4 at:
http://www.embedded.com/design/real-time-and-
performance/4007218/Digital-Signal-Processing-Tricks—High-speed-vector-
magnitude-approximation

URL (entegro). Sample clock offset detection and correction in the LTE
downlink. Accessed 2016 June 4 at: http://www.entegra.co.uk/wp-
content/uploads/2014/05/JOSP_OFDM_Sample_Clock_Offset.pdf

URL (EPC-Arch). Evolved Packet System. Accessed 06/12/2016 at:
http://read.pudn.com/downloads161/ebook/729955/Evolved%20Packet%20
System.pdf

URL (ethertronics1). Active antenna systems. Accessed 2016 June 4 at:
http://www.ethertronics.com/products/active-antenna-systems/

URL (ethertronics2). Cellular. Accessed 2016 June 4 at:
http://www.ethertronics.com/products/cellular/

URL (ETPS-LM3290). LM3290 product brief. Accessed 2016 June 4 at:

http://www.ti.com/lit/ds/symlink/lm3290.pdf

URL (eventhelix1). 3GPP LTE packet data convergence protocol (PDCP) sublayer. Accessed 2016 June 4 at: http://www.eventhelix.com/lte/presentations/3GPP-LTE-PDCP.pdf.

URL (eventhelix2). 3GPP LTE radio link control (RLC) sub layer. Accessed 2016 June 4 at: https://www.eventhelix.com/lte/presentations/3GPP-LTE-RLC.pdf.

URL (eventhelix3). 3GPP LTE channels and MAC layer. Accessed 2016 June 4 at: https://www.eventhelix.com/lte/presentations/3GPP-LTE-MAC.pdf.

URL (Farrow1). Fractional delay filters using farrow structures. Accessed 2016 June 4 at: http://www.mathworks.com/help/dsp/examples/fractional-delay-filters-using-farrow-structures.html

URL (free). LTE frequency bands. Accessed 2016 June 4 at: http://niviuk.free.fr/lte_band.php

URL (Fujitsu). MB86L13A LTE-optimized transceiver ideal for 4G LTE mobile handsets. Accessed 2016 June 4 at: http://www.fujitsu.com/downloads/MICRO/fswp/pdf/products/FSWP_RFT_MB86L13A_FS.pdf

URL (general dynamics). General dynamics LTE technology achieves distance milestone for long-range network communications. Accessed 2016 June 4 at: http://www.generaldynamics.com/news/press-releases/2015/06/general-dynamics-lte-technology-achieves-distance-milestone-long-range

URL (gps-gov). Telecom requirements for time and frequency synchronization. Accessed 2016 June 4 at: http://www.gps.gov/cgsic/meetings/2012/weiss1.pdf

URL (hyperphysics). Storing energy in a capacitor. Accessed 2016 June 4 at: http://hyperphysics.phy-astr.gsu.edu/hbase/electric/capeng2.html#c4

URL (IJWMN). SINR, RSRP, RSSI and RSRQ measurements in long term evolution. Accessed 2016 June 4 at: http://airccse.org/journal/jwmn/7415ijwmn09.pdf

URL (JESD207). Radio front end–baseband parallel (RBDP) interface. Accessed 2016 June 4 at: http://www.jedec.org/standards-documents/results/jesd207

URL (JOHANSON). RF antennas. Accessed 2016 June 4 at: http://www.johansontechnology.com/antennas

URL (jos). A wideband current-commutating passive mixer for multi-standard receivers in a 0.18 μm CMOS. Accessed 2016 June 4 at:

http://www.jos.ac.cn/bdtxben/ch/reader/create_pdf.aspx?file_no=12061903

URL (Keysight-Analysis). Signal analysis portfolio. Accessed 2016 June 4 at: http://www.keysight.com/upload/cmc_upload/All/ADDITIONAL-MATERIAL_Portfolio.pdf?&cc=US&lc=eng

URL (Keysight-EVM). Using error vector magnitude measurements to analyze and troubleshoot vector-modulated signals. Accessed 2016 June 4 at: http://cp.literature.agilent.com/litweb/pdf/5965-2898E.pdf

URL (Keysight-John). John Hansen, Vector modulation and frequency conversion fundamentals, Agilent Technologies. Accessed 2016 June 4 at: http://www.keysight.com/upload/cmc_upload/All/18July2013WebcastSlides.pdf

URL (Keysight-LTE). Long Term Evolution – LTE test. Accessed 2016 June 4 at: http://www.keysight.com/main/application.jspx?id=1174746&cc=US&lc=eng&cmpid=20331

URL (Keysight-MXG). MXG X-Series Signal Generators N5181B Analog & N5182B Vector. Accessed 2016 June 4 at: http://www.keysight.com/en/pcx-x205194/x-series-signal-generators-mxg-exg?cc=US&lc=eng

URL (Keysight-PN). Phase noise x-series measurement application. Accessed 2016 June 4 at: http://cp.literature.agilent.com/litweb/pdf/5989-5354EN.pdf

URL (Keysight-PXA). PXA X-series signal analyzer N9030A. Accessed 2016 June 4 at: http://www.keysight.com/en/pdx-x201775-pn-N9030A/pxa-signal-analyzer-3-hz-to-50-ghz?cc=US

URL (Keysight-VSA). 89600 VSA software. Accessed 2016 June 4 at: http://www.keysight.com/en/pc-1905089/89600-VSA-and-WLA-Software?cc=US&lc=eng

URL (limemicro1). Field programmable RF ICs: LMS7002M, features summary. Accessed 2016 June 4 at: http://www.limemicro.com/products/field-programmable-rf-ics-lms7002m/#features-summary

URL (linear-Synth). Uncompromising fractional-N synthesizers. Accessed 2016 June 4 at: http://cds.linear.com/docs/en/product-selector-card/Flyer_2PB_6948a.pdf

URL (litepoint1). Testing LTE - Where to begin? optimizing LTE test for IQxstream. Accessed 2016 June 4 at: http://www.litepoint.com/wp-content/uploads/2015/05/Testing-LTE_WhitePaper.pdf

URL (litepoint2). IQxstream® Multi-DUT cellular test system. Accessed 2016 June 4 at: http://www.litepoint.com/wp-content/uploads/2014/07/LP_IQxstream_Brochure_111513.pdf

URL (lmk). LTE channel coding and link adaptation. Accessed 2016 June 4 at: http://www.lmk.lnt.de/fileadmin/Lehre/Seminar09/Ausarbeitungen/Ausarbeitung_Zarei.pdf

URL (LMS7002M). FPRF MIMO transceiver IC with integrated microcontroller. Accessed 2016 June 4 at: http://www.limemicro.com/wp-content/uploads/2015/09/LMS7002M-Data-Sheet-v2.8.0.pdf

URL (lstelcom). Active antenna systems benefits and challenges. Accessed 2016 June 4 at: http://www.lstelcom.com/fileadmin/content/events/ls_summit_13_presentations/7_Presentation_LS_Summit_Chris_Haslett.pdf

URL (LTC-Balancer). LTC3128 – 3A monolithic buck-boost supercapacitor charger and balancer with accurate input current limit. Accessed 2016 June 4 at: http://www.linear.com/product/LTC3128

URL (LTE-Cat). LTE UE category & class definitions. Accessed 2016 June 4 at: http://www.radio-electronics.com/info/cellulartelecomms/lte-long-term-evolution/ue-category-categories-classes.php

URL (LTE-CatM). Category M (LTE-M). Accessed 2016 June 4 at: http://www.sharetechnote.com/html/Handbook_LTE_CategoryM.html

URL (LTE-Wiley). Appendix C, Deciphering the 3GPP LTE specifications. Accessed 2016 June 4 at: http://onlinelibrary.wiley.com/doi/10.1002/9780470758717.app3/pdf

URL (Mathworks-IQ). I/Q imbalance. Accessed 2016 June 4 at: http://www.mathworks.com/help/comm/ref/iqimbalance.html

URL (Mathworks-PSS). Synchronization signals (PSS and SSS). Accessed 2016 June 4 at: http://www.mathworks.com/help/lte/ug/synchronization-signals-pss-and-sss.html

URL (MAX2580). LTE small-cell MIMO transceiver. Accessed 2016 June 4 at: https://www.maximintegrated.com/en/products/comms/wireless-rf/MAX2580.html

URL (Max2837). MAX2837, 2.3GHz to 2.7GHz wireless broadband RF transceiver. Accessed 2016 June 4 at: http://datasheets.maximintegrated.com/en/ds/MAX2837.pdf

URL (MB8613L13A). MB86L13A LTE-optimized transceiver ideal for 4G LTE mobile handsets. Accessed 2016 June 4 at: http://www.fujitsu.com/downloads/MICRO/fswp/pdf/products/FSWP_RFT_MB86L13A_FS.pdf

URL (microchip). Crystal oscillator basics and crystal selection. Accessed 2016 June 4 at: http://ww1.microchip.com/downloads/en/AppNotes/00826a.pdf

URL (MIPI-Bat). Battery interface specification. Accessed 2016 June 4 at: http://mipi.org/specifications/battery-interface

URL (MIPI-DigRFv4). DigRF (SM) specifications. Accessed 2016 June 4 at: http://mipi.org/specifications/digrfsm-specifications#v4

URL (MIPI-eTrak). Analog control interface WG eTrak specification. Accessed 2016 June 4 at: http://mipi.org/specifications/analog-control-interface-wg-etrak%C2%AE-specification

URL (MIPI-HSI). High-speed synchronous serial interface (HIS) specification. Accessed 2016 June 4 at: http://mipi.org/specifications/high-speed-synchronous-serial-interface-hsi

URL (MIPI-LLI). Low latency interfaced specification. Accessed 2016 June 4 at: http://mipi.org/specifications/low-latency-interface-specification

URL (MIPI-RFFE). RF front-end specification. Accessed 2016 June 4 at: http://mipi.org/specifications/rf-front-end

URL (MIPI-SPMI). Introduction to RF front-end device communication. Accessed 2016 June 4 at: http://www.ni.com/white-paper/14355/en/

URL (MIPI-summary). MIPI interface in a mobile platform. Accessed 2016 June 4 at: http://mipi.org/about-mipi/mipi-interfaces-mobile-platform

URL (MIT-Synth). Sigma-Delta fractional-N frequency synthesis. Accessed 2016 June 4 at: http://www.ece.ucsb.edu/Faculty/rodwell/Classes/ece218b/notes/MTT_2004_meninger_IntNFracN.pdf

URL (NASA). The effect of Doppler frequency shift, frequency offset of the local oscillators, and phase noise on the performance of coherent OFDM receivers. Accessed 2016 June 4 at: http://ntrs.nasa.gov/archive/nasa/casi.ntrs.nasa.gov/20010049375.pdf

URL (nctu). Carrier frequency offset estimation algorithm for the 3GPP-LTE. Accessed 2016 June 4 at: http://diamondprj.nctu.edu.tw/files/842_9814045_-_doc.pdf

URL (NI-PA). Advanced PA test techniques digital predistortion and envelope tracking. Accessed 2016 June 4 at: ftp://ftp.ni.com/pub/branches/uk/ats_2014/Power_Amplifier_Testing.pdf

URL (nist1). Characterization of frequency and phase noise. Accessed 2016 June 4 at: http://tf.nist.gov/general/tn1337/Tn162.pdf

URL (nit). Normalized Gaussian approach to statistical modeling of OFDM signals. Accessed 2016 June 4 at: http://www.nit.eu/czasopisma/JTIT/2014/1/54.pdf

URL (NI-testing). Introduction to LTE device testing. Accessed 2016 June 4 at: http://download.ni.com/evaluation/rf/Introduction_to_LTE_Device_Testing.pdf

URL (noise). Noise: frequently asked questions. Accessed 2016 June 4 at: http://www.noisewave.com/faq.pdf

URL (noisecom). Noise: Terms and application webinar. Accessed 2016 June 4 at: http://noisecom.com/resource-library/webinars/noise-terms-and-applications-webinar

URL (Nokia). LTE-M - Optimizing LTE for the internet of things.

Accessed 2016 June 4 at: http://networks.nokia.com/file/34496/lte-m-optimizing-lte-for-the-internet-of-things

URL (nxp1). Long Term Evolution protocol overview. Accessed 2016 June 4 at: https://www.nxp.com/files/wireless_comm/doc/white_paper/LTEPTCLOVWWP.pdf

URL (ODI-Verizon). Open development device certification process. Accessed 2016 June 4 at: https://odi-device.verizonwireless.com/Info/Open%20Development%20Device%20Docs/Certification%20Process%20Documentation/ODDeviceCertificationProcess.pdf

URL (pscr). Extended cell testing. Accessed 2016 June 4 at: http://www.pscr.gov/projects/testing_evaluation/extended_cell/pscr_extended_cell_coverage.pdf

URL (PXA). N9030A PXA signal analyzer, 3 Hz to 50 GHz. Accessed 2016 June 4 at: http://www.keysight.com/en/pd-1721037-pn-N9030A/pxa-signal-analyzer-3-hz-to-50-ghz?cc=US&lc=eng

URL (Radisys). Protocol signaling procedures in LTE. Accessed 2016 June 4 at: http://go.radisys.com/rs/radisys/images/paper-lte-protocol-signaling.pdf

URL (Radisys-RAT). Interoperable UE handovers in LTE. Accessed 2016 June 4 at: http://go.radisys.com/rs/radisys/images/paper-lte-interoperable.pdf

URL (Rakon1). Relationship between phase noise and jitter. Accessed 2016 June 4 at: http://www.rakon.com/products/technical-resources/application-notes?start=5

URL (Rakon2). Phase noise and jitter in crystal oscillators. Accessed 2016 June 4 at: http://www.rakon.com/products/technical-resources/application-notes?start=5

URL (Rakon-OCXO). Ultra low noise OCXO. Accessed 2016 June 4 at: http://www.rakon.com/products/families/ocxo-ocso

URL (rfcafe). RMS and average voltage. Accessed 2016 June 4 at: http://www.rfcafe.com/references/electrical/sinewave-voltage-conversion.htm

URL (RFwireless1). LTE EPC network interfaces-S1 MME, S1-U, S3, S4, S5, S6a, Gx, S11, SGi. Accessed 2016 June 4 at: http://www.rfwireless-world.com/Tutorials/LTE-EPC-network-interfaces.html

URL (R&S-1). LTE UE RF measurements-An introduction and overview. Accessed 2016 June 4 at: https://www.rohde-schwarz.co.jp/events/webinar/lte_ue_rf_measurements_v122.pdf

URL (R&S-2). LTE system specifications and their impact on RF & base band circuits. Accessed 2016 June 4 at: http://cdn.rohde-schwarz.com/pws/dl_downloads/dl_application/application_notes/1ma221/1MA221_1e_LTE_system_specifications.pdf

URL (R&S-CMW500). R&S CMW 500 – platform overview. Accessed

2016 June 4 at: https://www.rohde-schwarz.com/us/product/cmw500_overview-powerslave_63491-10844.html

URL (R&S-LTE). LTE/LTE Advanced featured products. Accessed 2016 June 4 at: https://www.rohde-schwarz.com/au/solutions/wireless-communications/lte/products/products_53604.html

URL (Sanjole). Wavejudge 5000 LTE analyzer. Accessed 2016 June 4 at: http://www.sanjole.com/our-products/lte-analyzer/

URL (Skyworks). Basics of dual fractional-N synthesizers/PLLs. Accessed 2016 June 4 at: http://www.skyworksinc.com/uploads/documents/101463B.pdf

URL (Slideshare). LTE physical-layer overview. Accessed 2016 June 4 at: http://www.slideshare.net/praveenkmr78/lte-physical-layer-overview

URL (Spirent). VR5 overview. Accessed 2016 June 4 at: http://www.spirent.com/Products/VR5

URL (Symmetricom). Timing and synchronization for LTE-TDD and LTE-advanced mobile networks. Accessed 2016 June 4 at: https://www.aventasinc.com/whitepapers/WP-Timing-Sync-LTE-SEC.pdf

URL (taoglas-external). External 4G LTE antenna. Accessed 2016 June 4 at: http://www.taoglas.com/product-category/lte-antennas/external-lte/

URL (taoglas-internal). Internal 4G LTE antenna. Accessed 2016 June 4 at: http://www.taoglas.com/product-category/lte-antennas/internal-lte/

URL (tech-invite). 3GPP technical bodies. Accessed 2016 June 4 at: http://www.tech-invite.com/3m0/tinv-3gpp-ran.html

URL (ti-Synth). Fractional/integer-N PLL basics. Accessed 2016 June 4 at: http://www.ti.com/lit/an/swra029/swra029.pdf

URL (triquint1). B7 / B30 / B38 / B40 / B41N HB front-end module (FEM) with integrated B40 and B41 low drift BAW. Accessed 2016 June 4 at: http://www.triquint.com/products/p/TQF6297H

URL (UE-Cat). LTE ue-Category. Accessed 2016 June 4 at: http://www.3gpp.org/keywords-acronyms/1612-ue-category

URL (Ultracaps). Ultracapacitor and supercapacitor products. Accessed 2016 June 4 at: https://www.tecategroup.com/ultracapacitors-supercapacitors/cap-xx.php

URL (ursi). A miniature dielectric ceramics resonator antenna for mobile handset application. Accessed 2016 June 4 at: http://ursi.org/Proceedings/ProcGA02/papers/p1092.pdf

URL (Vectron-TCXO). Tutorial on TCXOs. Accessed 2016 June 4 at: http://www.vectron.com/products/literature_library/tutorial_on_tcxos.pdf

Glossary

3GPP	3rd Generation Partnership Project
A	Ampere, units of current
ACK	positive ACKnowledgement
ACLR	Adjacent Channel Leakage Ratio
ACS	Adjacent Channel Selectivity
ADC	Analog to Digital Converter
AGC	Automatic Gain Control
AMC	Adaptive Modulation and Coding
A-MPR	Additional Maximum Power Reduction
AP	Application Protocol
ARQ	Automatic Repeat reQuest
AS	Access Stratum
BB	Baseband
BBIC	BB Integrated Circuit
BOM	Bill Of Material
bps	billion parts per second
BS	Base Station (called as eNodeB in LTE)
BW	Band Width
Cat	Category, LTE Category
CCDF	Complementary Cumulative Distribution Function
CDF	Cumulative Distribution Function
Ceff	effective Capacitance
CF	Crest Factor
CFI	Control Format Indicator
CFO	Carrier Frequency Offset
CID	Cell IDentifier
Cos/cos	Cosine waveform
CP	Cyclic Prefix
CPICH	Common PIlot CHannel
CQI	Channel Quality Indicator
CRC	Cyclic Redundancy Check
CW	Continuous Wave
DAC	Digital to Analog Converter
dB	deciBel
dBm	deciBel power with 1 mW reference
DC	Direct Current (fixed or steady signal of current or voltage)
DFT	Discrete Fourier Transform
DL	Down Link or Downlink

DigRFv4	Digital RF version 4 interface goes with MIPI
DMRS	DeModulation Reference Signal
DRX	Discontinuous Reception
DSP	Digital Signal Processor
DTX	Discontinuous Transmission
DwPTS	Downlink Pilot Time Slot
EARFCN	E-UTRA Absolute Radio Frequency Channel Number
eDRX	extended Discontinuous Reception
eNB/eNodeB	evolved NodeB, base station in LTE
EPS	Evolved Packet System
ESR	Equivalent Series Resistor
ET	Envelope Tracking
ETSI	European Telecommunications Standards Institute
E-UTRA	Evolved Universal Terrestrial Radio Access, LTE radio access
EVM	Error Vector Magnitude
F	Farad, capacitance value
FDD	Frequency Division Duplex
FDMA	Frequency Division Multiple Access
FEM	Front End Module
FFT	Fast Fourier Transform
f_{vco}	VCO frequency
fW	Femto Watt 1×10^{15} Watts
GP	Guard Period
GPIO	General Purpose Input-Output
GPS	Global Positioning System
GW	GateWay, naming used in P-GW and S-GW in LTE
HARQ	Hybrid ARQ
HSI	High Speed synchronous serial Interface (MIPI-family)
HSS	Home Subscriber Server
Hz	Hertz, frequency in cycles per second
ICI	Inter-Carrier Interference
ID	Identity
IF	Intermediate Frequency
IFFT	Inverse Fast Fourier Transform
IFO	Integer Frequency Offset
IM2/IMD2	Inter Modulation Distortion 2
IM3/IMD3	Inter Modulation Distortion 3
IoT	Internet-of-Things
IP	Internet Protocol
IQ/(I-Q)	In-phase and Quadrature-phase
ITU	International Telecommunication Union
ITU-R	International Telecommunication Union Radio

	communication sector
Joules	Unit of energy (power × time duration)
k_B	Boltzmann constant (used with noise power calculation)
kHz	kilo Hz (1000 cycles per second)
km/h	kilo meters per hour
L1	Layer-1
L2	Layer-2
L3	Layer-3
LDO	Low DropOut voltage regulator
LLI	Low Latency Interface
LNA	Low Noise Amplifier
LO	Local Oscillator
LOFT	LO Feed-Through (or leakage)
LOS	Line-Of-Sight
LPF	Low Pass Filter
LTE	Long-Term Evolution
LTE-M	short name for LTE-MTC
LTE-MTC	LTE Machine Type Communications
M2M	Machine to Machine
mA	milli Amperes (1/1000th of Ampere)
MAC	Medium Access Control
MBMS	Multimedia Broadcast and Multicast Services
MCS	Modulation and Coding Scheme
MHz	Mega Hz (million cycles per second)
MIB	Master Information Block
milli	1/1000th
MIMO	Multiple-Input Multiple-Output
MIPI	Mobile Industry Processor Interface
MME	Mobility Management Entity
MPR	Maximum Power Reduction
ms	milli seconds (1/1000th of second)
MTC	Machine Type Communication
mV	milli Volt (1/1000th of Volt)
mW	milli Watt (1/1000th of Watt)
NACK	Negative ACKnowledgment
NAS	Non-Access Stratum
NB	Narrow-Band
NB-IoT	Narrow-Band Internet-of-Things
NF	Noise Figure
NS	Network Signaling
ns	nano seconds (10^{-9} seconds)
OCXO	Oven Compensated XO
OFDM	Orthogonal Frequency Division Multiplexing

OOB	Out Of Band
P1dB	1 dB compression point
PA	Power Amplifier
PAPR	Peak-to-Average Power Ratio
PAR	Peak-to-Average Ratio
PBCH	Physical Broadcast CHannel
PCB	Printed Circuit Board
PCCH	Paging Control CHannel
PCFICH	Physical Control Format Indicator CHannel
PCH	Paging CHannel
PCI/PCID	Physical Cell Identifier
PCRF	Policy and Charging Rules Function
PDCCH	Physical Downlink Control CHannel
PDCP	Packet Data Convergence Protocol
PDF/pdf	Probability Distribution Function
PDN	Packet Data Network
PDP	Power Delay Profile
PDSCH	Physical Downlink Shared CHannel
PDU	Protocol Data Unit
PFD	Phase Frequency Detector
P-GW	Packet Data Network GateWay
PH	Power Headroom
PHICH	Physical Hybrid ARQ Indicator CHannel
PHY/Phy	Physical layer
Pk-Pk	Peak to Peak (goes with voltage)
PL	Path Loss
PLL	Phase Locked Loop
PLR	Packet Loss Rate
PMCH	Physical Multicast CHannel
PMI	Precoding Matrix Indicator
PN	Pseudo Noise
PPB/ppb	Parts Per Billion
PPM/ppm	Parts Per Million
PRACH	Physical Random Access CHannel
PRB	Physical Resource Block, 12 REs in one LTE slot
$P_{REFSENS}$	REFerence SENSitivity Power (level)
PRN	Pseudo Random Noise
ps	pico seconds (10^{-12} seconds)
PSM	Power Saving Mode
PSS	Primary Synchronization Signal
pW	pico Watt
PUCCH	Physical Uplink Control CHannel
PUSCH	Physical Uplink Shared CHannel

PVT	Power Voltage Temperature
QAM	Quadrature Amplitude Modulation
QPSK	Quadrature Phase Shift Keying
RA	Random Access
RACH	Random Access CHannel
RAN	Radio Access Network
RAR	Random Access Response
RAT	Radio Access Technology
RB	Resource Block
RE	Resource Element, one LTE tone
REFSENS	REFerence SENSitivity power level
RF	Radio Frequency
RFFE	RF Front End, RF MIPI interface
RF FEM	RF Front End Module
RFIC	RF Integrated Circuit
RFO	Residual Frequency Offset
RLC	Radio Link Control
RMS/rms	Root Mean Square
RRC	Radio Resource Control, Layer-3 control plane
RRM	Radio Resource Management
RS	Reference Signal
RSRP	Reference Signal Received Power
RSRQ	Reference Signal Received Quality
RSSI	Received Signal Strength Indicator
RTC	Real Time Clock
RV	Redundancy Version
Rx/RX	Receive or Receiver
RxD	Receiver Diversity
s	seconds
S1	The eNodeB and the Core Network (CN) interface
S1-AP	S1 Application Protocol
S1-U	S1-User plane
S5	interface between a S-GW and P-GW in same network
S8	interface between S-GW and P-GW in different network
SAW	Surface Acoustic Wave
SC-FDMA	Single-Carrier Frequency Division Multiple Access
SCO	Sampling Clock Offset
SCTP	Streaming Control Transmission Protocol
SDU	Service Data Unit
SEM	Spectrum Emission Mask
SF	Sub-frame or SubFrame
S-GW/SGW	Serving GateWay
SI	System Information

SIB	System Information Block
Sin/sin/sine	Sinusoidal waveform
SINR	Signal to Interference plus Noise Ratio
SNR	Signal to Noise Ratio
SPI	Serial Peripheral Interface
SPMI	System Power Management Interface
SRS	Sounding Reference Signals
SSF	Special Sub Frame (in LTE TDD)
SSS	Secondary Synchronization Signal
STD	STandard Deviation
Supercap	Super Capacitor
T	Temperature in this book
TA	Timing Advance
TAS	Time alignment Strobe goes with MIPI
TCP	Transmission Control Protocol
TCXO	Temperature Compensated XO
TDD	Time Division Duplex
TDM	Time Division Multiplexing
TDMA	Time Division Multiple Access
TIA	Trans Impedance Amplifier
TPC	Transmitter Power Control
Ts	LTE timing resolution 1/30.72 MHz or 32.552 ns
Tx/TX	Transmit or Transmitter
UCI	Uplink Control Information
UDP	User Datagram Protocol
UE	User Equipment
UL	Up Link or Uplink
UM	Unacknowledged Mode
UMTS	Universal Mobile Telecommunications System
UpPTS	Uplink Pilot Time Slot, applicable in LTE TDD
UTRA	Universal Terrestrial Radio Access
UTRAN	Universal Terrestrial Radio Access Network
Uu	interface between eNodeB and UE
V	Volts, units of voltage
V_{BAT}	Power amplifier supply voltage
VCO	Voltage Controlled Oscillator
VIO	Voltage Input Output (IO interface voltage)
VoIP	Voice over Internet Protocol
VoLTE	Voice over LTE
V_r	Velocity
Vrms	Voltage rms (root means square) value
W	Watt, units of power
WG	Working Group (goes with RAN)

X2	interface to interconnect eNodeBs
XO	Crystal Oscillator
ZC	Zadoff-Chu sequence
μF	micro Farad (10^{-6} Farads)
μs	micro second (10^{-6} seconds)
μW	micro Watt (10^{-6} Watts)
Ω	Ohms, units of impedance or resistance
σ	Symbol used for STD in this book, σ^2 is variance

www.ingramcontent.com/pod-product-compliance
Lightning Source LLC
Chambersburg PA
CBHW071813200526
45169CB00017B/209